EXPERT SYSTEMS
Applications for Structural, Transportation, and Environmental Engineering

M. Arockiasamy
Professor of Ocean Engineering
Department of Engineering
Florida Atlantic University
Boca Raton, Florida

CRC Press
Boca Raton Ann Arbor London Tokyo

Library of Congress Cataloging-in-Publication Data

Arockiasamy, M.
Expert systems application for structural, transportation, and environmental engineering / author, M. Arockiasamy.
p. cm.
Includes bibliographical references and index.
ISBN 0-8493-4217-1
1. Civil engineering--Data processing. 2. Expert systems (Computer science) 3. Structural engineering--Data processing. 4. Transportation engineering--Data processing. 5. Environmental engineering--Data processing. I. Title.
TA345.A76 1992
624'.0285'633--dc20 92-18992
CIP

This book represents information obtained from authentic and highly regarded sources. Reprinted material is quoted with permission, and sources are indicated. A wide variety of references are listed. Every reasonable effort has been made to give reliable data and information, but the author and the publisher cannot assume responsibility for the validity of all materials or for the consequences of their use.

Direct all inquiries to CRC Press, Inc. 2000 Corporate Blvd., N.W., Boca Raton, Florida, 33431.

International Standard Book Number 0-8493-4217-1

Library of Congress Card Number 92-18992

Printed in the United States of America 1 2 3 4 5 6 7 8 9 0

Printed on acid-free paper

PREFACE

Knowledge-based-expert systems is that branch of Artificial Intelligence (AI) that deals specifically with automating the advice-giving capabilities of human experts. The expert systems have gained popularity recently with the sharp decrease in the cost of information technology. There is a large number of expert systems development activities in almost all fields including architecture, engineering and science.

This book provides a wide coverage of the basic concepts of expert systems together with the applications in the areas of structural, construction, transportation, and environmental engineering. The material and scope are directed primarily to practicing architects, engineers, system designers, programmers, instructors in colleges and universities; in addition, the book will serve as a text for senior undergraduate and graduate students to increase awareness and understanding of the versatility of the expert systems in civil engineering.

The book is structured in seven chapters with a comprehensive presentation of expert system applications. Chapter 1 has been designed as an introduction to expert systems for readers who are new to the field. It describes the expert system components including the characteristics of expert systems. Chapter 2 presents the expert system methodologies for design applications, and selected applications to structural design — preliminary design of three-dimensional grid, design system for low-rise industrial buildings, preliminary design of frameworks, bridge design system, and retaining wall design. Chapter 3 covers design standards to improve the representation, organization, analysis and use of standards. Chapter 4 outlines typical expert systems intended for construction engineering and management applications, namely, soil exploration consultant, layout of temporary construction facilities, site layout expert system, construction planning, construction scheduling knowledge representation, and maintenance.

Chapter 5 includes the main concepts underlying the expert systems with emphasis on bridge analysis, rating, and management. Chapter 6 describes expert systems methodology and applications which aid the transportation and highway engineer in planning, design, and operation. Chapter 7 focuses on several applications of expert systems in the fields of environmental and water resources engineering. Typical prototype expert systems are intended for hazardous site evaluation, combatting oil spill pollution, resolution of air quality and reservoir analysis.

The material covered in this book serves as an introduction to the field of knowledge-based expert systems with applications to civil engineering. Although several applications are currently under development, this book provides the reader with the flavor, concepts, and content of current expert system technology and applications. The types of problems and the solution techniques are in many cases common to different areas and, therefore, the reader will stand to gain more by perusing the whole book rather than just that chapter specific to his/her immediate area of interest.

A number of people contributed significantly to make this book possible. The contents of Chapters 1 through 4 have been developed based on the support provided by the U.S. Army Engineer Waterways Experiment Station (WES), Information Technology Laboratory (ITL), Vicksburg Mississippi. The Author gratefully acknowledges the comments of Dr. N. Radhakrishnan, Chief, (ITL) during the course of study.

The editorial assistance provided by Mr. Sunghoon Lee and Mr. Rajasekhar Kalyandurg, Graduate Assistants, is greatly appreciated. Thanks are due to the authors and publishers of many articles and books, listed in the bibliography, but too numerous to mention here, who kindly permitted the reproduction of some figures and tables in the various chapters.

Finally, I must record my gratitude to my wife, Glory, who provided the moral support to make this book possible.

M. Arockiasamy
Boca Raton, Florida

THE AUTHOR

M. Arockiasamy, Ph.D. (Wisconsin), M.Sc. and B.E. (Hons.) (Madras) is Professor of Ocean Engineering at Florida Atlantic University, Boca Raton, Florida, where he teaches several undergraduate and graduate courses in mechanics and structures. He is a registered professional engineer in Alabama, Florida, Wisconsin and Newfoundland with a long and distinguished research and teaching career.

Dr. Arockiasamy has taught at the University of Madras and Memorial University of Newfoundland. He has worked in industry and acted as consultant to a large number of companies. His research interests are in the areas of structural engineering and mechanics, computer aided design, knowledge based expert systems applied to design, construction and maintenance of hydraulic structures, selection and design of retaining structures and flood walls, and bridge rating and management, software engineering, marine structures and materials, static and cyclic behavior of prestressed concrete bridges, fatigue of offshore tubular welded joints, active control of deformations in bridges, and composite structures.

He has over 200 research and several technical publications, and is an active member of several technical committees of the American Concrete Institute, the American Society of Civil Engineers and the Transportation Research Board. He is also a co-author/editor of three other books on structural analysis, prestressed concrete, and offshore structures. He was recognized for excellence in teaching in 1992 by Florida Atlantic University.

TABLE OF CONTENTS

Chapter 1

INTRODUCTION

1.1. WHAT IS AN EXPERT SYSTEM?

Expert systems (ES) are emerging as a means of automating the solution to problems that have not yet been formalized as algorithms. Applications of ES range from medical diagnosis to architectural design. Although many tools are available for the development of ES that use classification or diagnostic problem-solving strategies, very few tools are available that provide an environment for the development of a hierarchical planning or design strategy. ES is a useful tool for solving ill-defined problems such as those in structural design, where intuition and experience are necessary ingredients. This section defines ES so as to establish a common vocabulary and a brief review of available tools.

ES are generally defined as interactive computer programs incorporating judgment, experience, rules of thumb, intuition, and other expertise to provide knowledgeable advice about a variety of tasks (Gaschnig et al., 1981; Fenves, 1986; Maher, 1987; Adeli, 1988). This definition does not clearly distinguish ES from traditional computer programs. The traditional programs can be interactive, and contain judgment and rules of thumb, yet they are not ES. The characterizing features of conventional programs and ES are listed in Table 1.

1.2. EXPERT SYSTEM ARCHITECTURE

Knowledge-based expert systems (KBES) have been identified, based on research in artificial intelligence, as practical problem-solving tools. The basic architecture of an ES has three basic components: the knowledge base, the context, and the inference mechanism. User interface and an explanation facility are two additional components that make the ES more usable. A knowledge acquisition facility also is desirable to enhance the extensibility of the ES. The components of an ES are shown in Figure 1 (M. Arockiasamy et al., 1989).

The knowledge base in the ES contains the facts and heuristics associated with the domain in which the ES is applied. The facts are typically represented as declarative knowledge, whereas heuristics take the form of rules. Modification of the knowledge base is important in most engineering domains, since knowledge is continually changing and expanding. Many ES environments provide higher-level representation schemes than procedural code, such as rules or frames to make the knowledge base as transparent as possible.

The context is the component of the ES that initially contains the information that defines the parameters of the problem. As the ES reasons about

TABLE 1
Characteristics of Traditional Programs and Expert Systems

Traditional programs	Expert systems
Representation and use of data	Representation and use of knowledge
Knowledge and control integrated	Knowledge and control separated
Algorithmic (repetitive) process	Heuristic (inferential) process
Effective manipulation of large databases	Effective manipulation of large knowledge bases
Programmer must ensure uniqueness and completeness	Knowledge engineer inevitably relaxes uniqueness and completeness restraint
Midrun explanation impossible	Midrun explanation desirable and achievable
Oriented toward numerical processing	Oriented toward symbolic processing

From Maher, M. L., Expert System Components, ASCE Expert Systems Comm. Rep. Expert Systems for Civil Engineers: Technology and Application, American Society of Civil Engineers, New York, 1987.

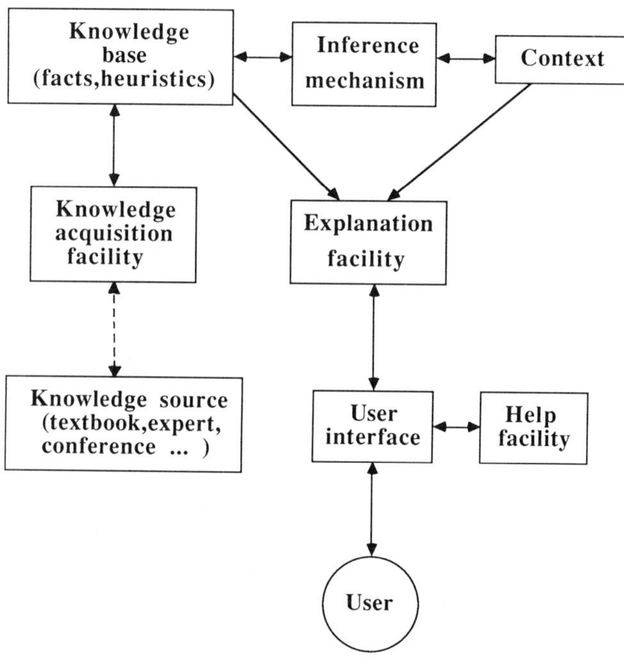

FIGURE 1. Components of an ES.

the given problem, the context expands to include the information generated by the ES to solve it. At completion of the problem-solving process, the context contains all the intermediate results of the problem-solving process as well as the solution. The context is a declarative form of the current state of the problem the ES is solving.

The inference mechanism contains the control information and uses the knowledge base to modify and expand the test. It controls the reasoning strategy of the ES through assertions, hypotheses, and conclusions. The reasoning process is controlled by the inference mechanism at different levels. When it operates at very low levels, providing flexibility in solution strategy, the knowledge base shall contain additional control information specific to the application domain. With a more specific inference mechanism, the control information will be less in the knowledge base.

The explanation facility in an ES provides answers to questions about the reasoning process used to develop a solution. A good explanation facility can explain both why a certain fact is requested and how a certain conclusion was reached. The knowledge acquisition facility in an ES is the component that facilitates the structuring and development of the knowledge base. This facility acts as an editor, and the expert should be able to add to or modify the knowledge base as, and when, the ES reveals gaps in the knowledge base. The knowledge acquisition facility understands the inference mechanism being used and can actively aid the expert in defining the knowledge base.

The user interface in the ES allows the traditional capabilities of conventional user interfaces. It allows the user to interact with and query the ES. In addition to being highly interactive, perhaps with "help" facilities, an ES user interface needs a transparency of dialogue, whereby some form of an explanation facility indicates the inference or reasoning process used.

1.3. ARCHITECTURAL VARIATIONS

The production system model and the blackboard model are two of the most common variations in the basic architecture. The production system represents a powerful model for human information processing and problem-solving ability. The blackboard model introduces the concept of multiple knowledge sources for handling complex problems.

1.3.1. PRODUCTION SYSTEM MODEL

The production system model considers the knowledge base as a set of rules, termed production memory. A production system consists of three main elements:

1. A set of IF-THEN rules or knowledge base
2. A global database or working memory
3. An inference mechanism

The rules are developed by the expert and need not be specified in the order in which they are to be considered. The inference mechanism in a production system provides the underlying strategy for identifying the productions that are eligible to be executed and the selection of one of these

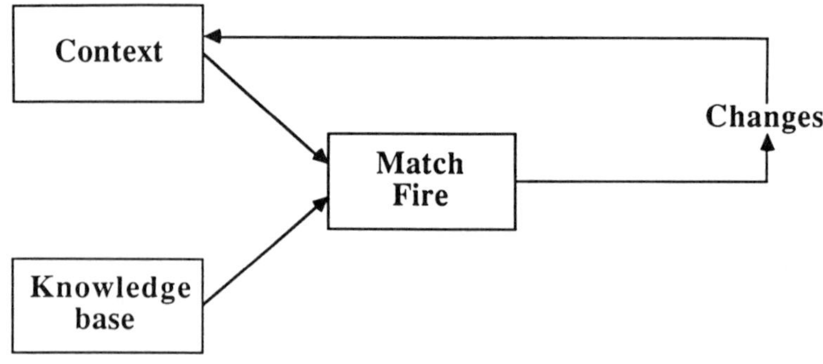

FIGURE 2. Production system model.

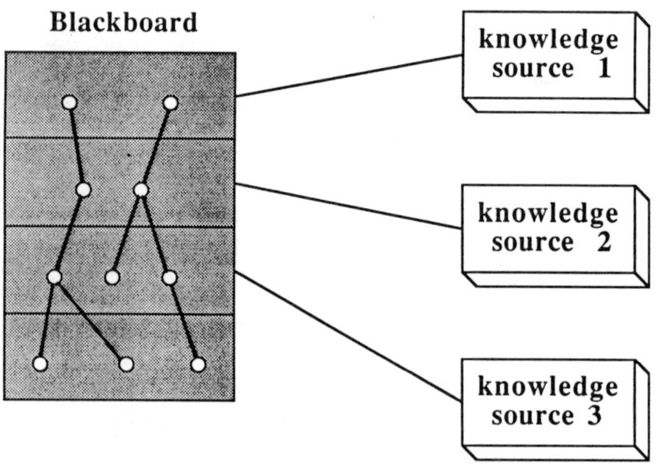

FIGURE 3. Blackboard model. (From Maher, M. L., *Expert System Components,* ASCE Expert Systems Comm. Rep. Expert Systems for Civil Engineers: Technology and Application, American Society of Civil Engineers, New York, 1987. With permission.)

productions. The inference mechanisms, viz., forward chaining and backward chaining, fire rules according to the built-in reasoning process. Figure 2 shows an illustrative production system model. The earliest implementations of the production system model (VanMelle, 1979) are EMYCIN and OPS5 (Forgy, 1981).

1.3.2. BLACKBOARD MODEL

The blackboard model illustrated in Figure 3 is based upon separation of the knowledge base into knowledge sources and the use of a blackboard as a context. The blackboard, a central global database, serves as a communication vehicle among knowledge sources and keeps track of incremental changes made in the current state of the problem until a solution is found. The blackboard utilizes a combination of forward- and backward-reasoning chains. The blackboard concept was first implemented in HEARSAY-II (Reddy et al., 1973). The blackboard model has been applied to problems involving distributed processing, multiple levels of knowledge, and multiple sources of knowledge. The problems being solved by the use of a blackboard model tend to be complex and, hence, require partitioning into subproblems.

1.4. PROGRAMMING LANGUAGES AND TOOLS FOR BUILDING EXPERT SYSTEMS

1.4.1. GENERAL PURPOSE PROGRAMMING LANGUAGES

Expert systems (ES) can be written in any programming language, such as LISP, PROLOG, C, FORTRAN, or PASCAL. LISP, which is still the choice of many developers in the U.S., was one of the first languages directed toward symbolic representation and list processing. The concept of structured programming incorporated into PASCAL reduces the complexity through modular programming and effective communication; it allows the programmer to define variable types such as character, string, boolean (with values of either true or false), integer, real number, and array. PASCAL has a variable-type pointer, which makes it possible to define logical trees. It can also be used for dynamic storage allocation. Turbo PASCAL has excellent string manipulation and powerful graphic capabilities. C is a very efficient language and is especially suitable for graphic-based programs. While LISP is memory intensive and requires large processing power, C has limited symbolic manipulation and memory management capabilities.

PROLOG (PROgramming LOGic), which is based on formal logic, is popular in Europe and Japan. It has its own inference mechanism. Experience with PROLOG-based ES shells shows that PROLOG is a versatile language for database-type applications (Allwood et al., 1985). However, certain limitations regarding numeric data types, large memory requirement, and slow execution with many implementations of the language are reported for ES development.

1.4.2. RESEARCH TOOLS

Selection of an ES shell for engineering applications should be based on type of application, type of machine and operating systems, maximum number of rules allowed (in production systems), response time (in solving problems or answering questions), type of control strategy and inference mechanism, user interface (graphics, natural language processing, etc.), availability of complex mathematical routines, ability to interface with other programs written in the language of the shell, programming aids (editors, debuggers, and a help facility), user support, etc. For engineering problems, numerical algorithmic routines must usually be combined with heuristics.

Although a number of ES have been developed, only a few of the more relevant ES tools are described below.

The first widely used ES shell was created by stripping the medical knowledge base from MYCIN, and called EMYCIN (for Essential MYCIN or Empty MYCIN), which is used to construct diagnosis systems. EMYCIN is LISP based and uses production rules that have the form of associative (object-attribute-value) triples for knowledge representation and backward chaining as the inference mechanism. It has been used to develop SACON (Structural Analysis CONsultant), an ES for the application of a general-purpose, finite element, structural analysis program (VanMelle, 1979; Bennett and Engelmore, 1979). PROSPECTOR also led to the development of another ES shell called KAS (Knowledge Acquisition System), which uses rule-based representation with a partitioned semantic net for organizing the process of rule matching. KAS, which was implemented in INTERLISP, uses both backward- and forward-chaining and certainty factors, and has explanation knowledge acquisition and tracing facilities (Reboh, 1981). EXPERT, which is a major ES shell implemented in FORTRAN, has explanation, knowledge acquisition, consistency checking, and trace facilities. When the ES developer adds a new rule, EXPERT tests the consistency of the rule with the solutions of the representative cases stored in the database. The framework of ES tools shown in Figure 4 can be used as comparative criteria to make the best choice of possible tools for a specific application.

1.5. KNOWLEDGE ELICITATION PROCESS (Firlej, 1985)

The real problems involved in building ES are those related to knowledge representation. The emphasis in the building of ES always seems to be on investigating technical issues and implementing the knowledge already elicited. The overall nature of the task is to extract knowledge from an expert in such a way as to reduce the risks and costs involved in the construction of a knowledge-based system. The information in the knowledge base of the ES can be obtained from two sources: literature and domain-specific knowledge from experts. Literature sources include technical journals, textbooks, manuals, public and commercial documents, and reports. A second source of domain-specific knowledge is from experts, who aid in the development of the system by providing their experience, intuition, judgment, rule of thumb, etc. Before contacting domain experts, the knowledge engineer (system developer) needs to review relevant literature in order to structure questions for the experts in such a way that the specific information sought is given naturally, without tension.

It is essential to avoid dislocations during the interview, e.g., to know when to keep quiet and when to prompt, when to direct and when to let the information flow. Since eliciting information from the expert on a large project

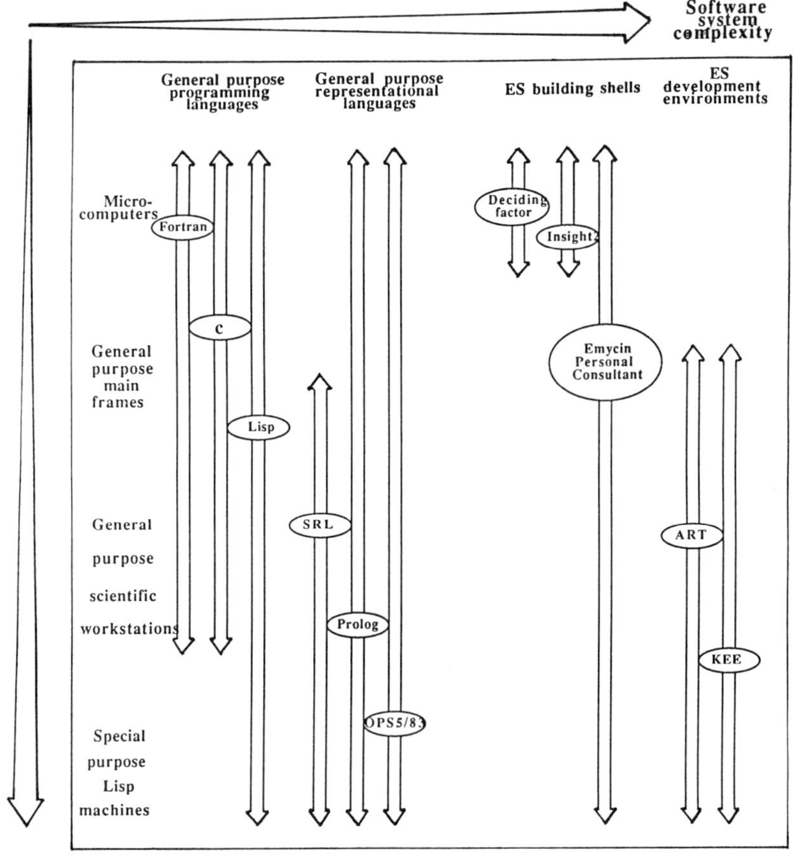

FIGURE 4. ES building-tool identification network. (From Maher, M. L., *Expert System Components,* ASCE Expert Systems Comm. Rep. Expert Systems for Civil Engineers: Technology and Application, American Society of Civil Engineers, New York, 1987. With permission.)

might take several years, it is essential that the expert's interest and motivation is maintained throughout that period. An expert who finds the whole process tiring and unpleasant will show his feelings in the quality of his response. The obstacles and problems must be identified well in advance so that the elicitation process can proceed without interruptions. Practical issues, such as the tape recording and transcribing of interviews, must be organized efficiently beforehand so that the analysis of information is not delayed unnecessarily.

Chapter 2

EXPERT SYSTEMS IN STRUCTURAL DESIGN

2.1. INTRODUCTION

An overview of expert systems (ES) in civil engineering is presented in recent references (Fenves et al., 1984; Maher, 1987; Adeli, 1988). Potential applications of artificial intelligence (AI) in structural engineering design and detailing were first proposed by Fenves and Norabhoompipat (1978). An expanded model of the design process was proposed by Rooney and Smith (1983) by introducing a feedback mechanism consisting of (1) acquisition of experience, (2) application of experience, and (3) database management. This model was then applied to a single-span, simply supported, steel wide-flange beam. Most ES developed thus far are basically experimental systems that show the present status and potential applications or present conceptual frameworks.

2.2. STRUCTURAL DESIGN PROCESS

The need to transmit loads in space to a support or foundation is first defined subject to constraints on cost, geometry, and other criteria. The design process finally yields the detailed specifications of a structural configuration that would transmit the given loads with the desired levels of safety and serviceability. The three sequential stages in the design process are preliminary design, analysis, and detailed design.

2.2.1. PRELIMINARY DESIGN

The conceptual design relates to the synthesis of potential configurations satisfying a few principal constraints. Synthesis of feasible structural configurations based on subsystems applicable to the particular design at hand, formulation and evaluation of specific constraints applicable to the chosen configurations, and choice of one or more of these configurations are the important aspects of the preliminary design stage.

2.2.2. ANALYSIS

This is the process of modeling the selected structural configuration and determining its response to external effects. Transformation of the real structural configuration to a mathematical model, selection and use of analysis procedures, and interpretation of analytical results in terms of the actual physical structure form the important components of this stage.

2.2.3. DETAILED DESIGN

This stage refers to the selection and proportioning of structural components that would satisfy all applicable constraints. This is again subdivided into a series of essentially hierarchical subproblems such as detailing the main structural components (beams, columns, etc.), followed by detailing of their subcomponents (connections, reinforcement, etc.). Within each subproblem, a further subdivision is made for selection based on certain controlling constraints (load-carrying capacity or buckling), followed by the evaluation of secondary constraints (e.g., local buckling or crippling).

A reanalysis would be required if the properties of components assumed at the analysis stage show significant deviations from those determined at the detailed design stage. Major and minor cycles of redesign may be necessary until a satisfactory optimal design is obtained. The *conceptualize-analyze-detail* cycle is characteristic of any design example (M. Arockiasamy et al., 1989).

2.3. EXPERT SYSTEM METHODOLOGIES FOR DESIGN

The derivation approach and the formation approach are the two basic methods used in expert systems (ES). The derivation approach involves deriving the most appropriate solution for the given problem from a list of predefined solutions stored in the knowledge base of ES, whereas the formation approach yields a solution from the eligible solution components stored in the knowledge base. An ES may use one or both of these approaches, depending on the complexity of the problem being solved.

The search for a solution of the problem solving using the formation approach begins at an initial state of known facts and conditions, which are combined to form a goal state. In a derivation approach, the known facts and conditions are used to derive the most appropriate goal state.

Forward chaining, backward chaining, and mixed initiative are appropriate strategies for the implementation of a derivation approach. The goal states represent the potential solutions and the initial state represents the input data. The development of an inference network representing the connections between initial states and goal states is illustrated in Figure 1. The advantage of using one of these strategies is that they are currently implemented in a variety of ES tools so that the development process involves defining, testing, and revising an inference network.

Problem reduction, plan-generate-test, and agenda control are problem-solving strategies appropriate for implementing a formation approach. The concepts of hierarchical planning and least commitment, backtracking, and constraint handling techniques could supplement these strategies. Figure 2 illustrates the unconnected graph of components. The solution is not completely defined by a goal state, but requires that the solution path also be known. The disadvantage of using one of these strategies is the lack of a

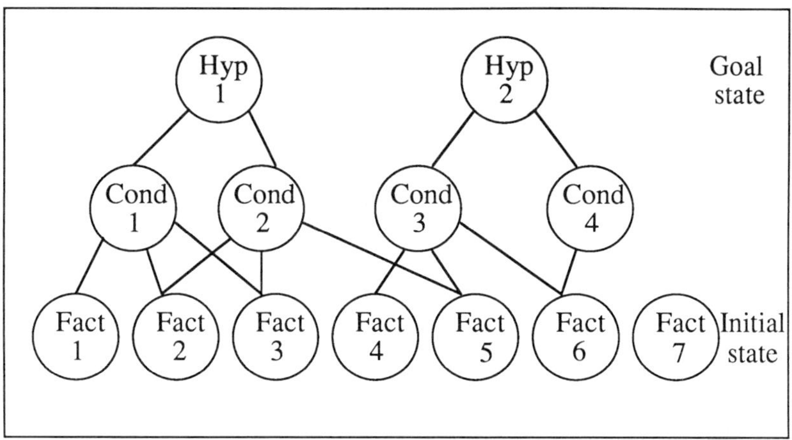

FIGURE 1. Inference network for a derivation problem. (From Maher, M. L., *Expert Systems in Civil Engineering,* American Society of Civil Engineers, New York, 1986, 7. With permission.)

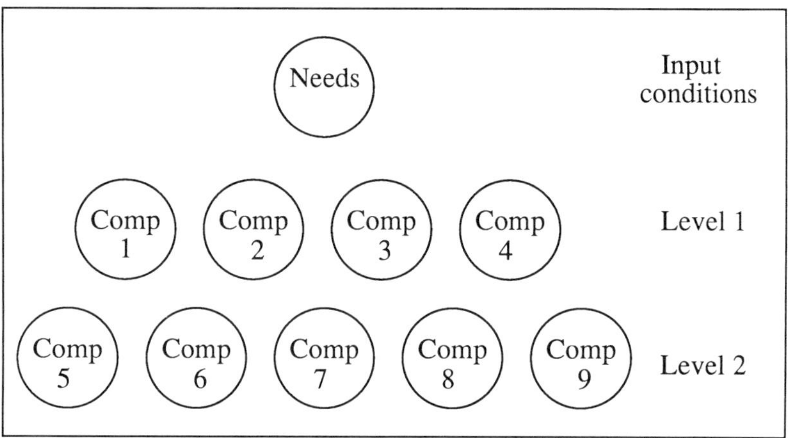

FIGURE 2. Unconnected graph for a formation problem. (From Maher, M. L., *Expert Systems in Civil Engineering,* American Society of Civil Engineers, New York, 1986, 7. With permission.)

standard implementation or ES tool that employs a strategy appropriate for the formation approach. These strategies are typically implemented using a lower-level language such as LISP or an ES shell such as KEE (Knowledge Engineering Environment).

Representation and use of constraints are essential in any design application. Three operations on constraints are proposed by Stefik (1980):

Constraint formulation—This operation adds new constraints representing restrictions on variable bindings.

TABLE 1
Expert Systems in Structural Design

System	Current state	Machine/developer
RETWALL	Developmental prototype	SUN 2
BDES	Developmental prototype	IBM PC
WISER	Developmental prototype	Symbolics 3640
HI-RISE	Developmental prototype	VAX 11/750
LOW-RISE	Operational prototype	VAX 11/750
ALL-RISE	Operational prototype	VAX 11/750
SFOLDER	Operational prototype	VAX 11/750
HI-COST	Operational prototype	VAX 11/750
DESTINY	Developmental	VAX 11/750
SSPG	Developmental	The Ohio State University
BTEXPERT	Prototype	The Ohio State University
RTEXPERT	Developmental	The Ohio State University
Preliminary design of frameworks by expert systems	Developmental	University of South Western Louisiana

Constraint propagation—This operation combines old constraints to form new constraints. It deals with interactions between subproblems through the reformulation of constraints from different subproblems.

Constraint satisfaction—This operation finds values for variables so that the constraints on these variables are satisfied.

Table 1 presents selected ES applications to structural design. Brief descriptions of specific applications of ES are described in the following sections. Each application is presented with a general description of the problem, the methodology employed, the current state of the system, and references.

2.4. EXPERT SYSTEM APPLICATIONS IN STRUCTURAL DESIGN

2.4.1. PRELIMINARY DESIGN: HI-RISE

The preliminary structural design is based on the designer's experience as well as on understanding of the behavior of structural systems. Outlining a structural system for a given building requires a combination of structural system knowledge, experience, and creativity. HI-RISE is an ES that forms and evaluates several alternative structural systems for a given three-dimensional grid. The expertise in HI-RISE is derived primarily from a recent publication on preliminary structural design (Lin and Stotesburg, 1981) using approximate analysis techniques and applicable design heuristics.

Classes of generic structural subsystems are used as a basis for the generation of feasible systems. Some examples of structural subsystems are rigidly connected frames, cores, trussed tubes, and braced frames. The generic structural subsystems are expanded and combined to fit the conditions of the

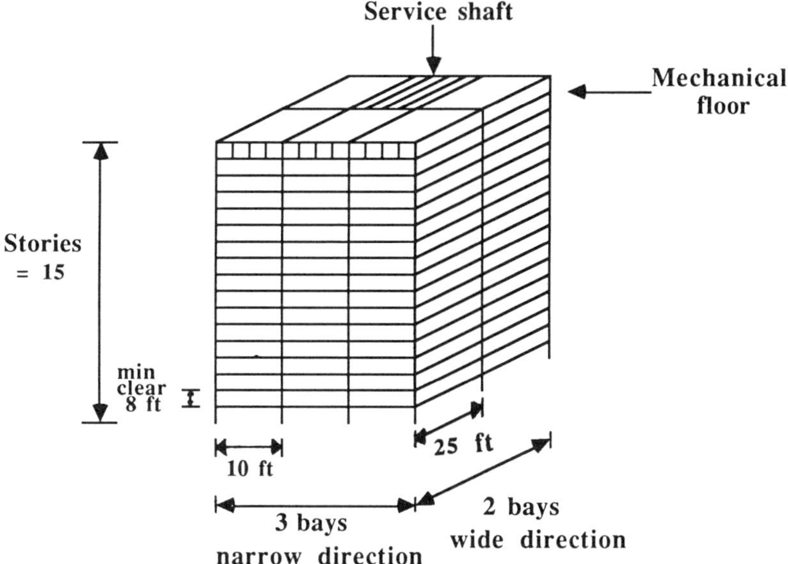

FIGURE 3. Graphical representation of input. (From Maher, M. L., HI-RISE: A Knowledge-Based Expert System for the Preliminary Design of High Rise Buildings, Ph.D. thesis, Carnegie-Mellon University, Pittsburgh, PA, 1984. With permission.)

particular building. HI-RISE was developed using the PSRL language running on a DEC VAX system. PSRL provides a combination of frame-based and rule-based reasoning. Frames are used in HI-RISE to represent the knowledge of structural systems, subsystems, and components in a hierarchical manner. Rules are used to represent strategy and heuristic knowledge. LISP functions are used to represent approximate analysis procedures.

HI-RISE decomposes the structural design process into five subtasks: synthesis, analysis, parameter selection, evaluation, and system selection. The synthesis subtask functions as a first search through the hierarchy of structural subsystems, using heuristic constraints to eliminate infeasible alternatives. The analysis subtask provides for an approximate analysis of a feasible alternative in order to determine the load distribution. The parameter selection subtask proportions key components. The evaluation subtask ranks all feasible alternatives using a heuristic evaluation function. System selection can be done by the user or by defaulting to the system with the best evaluation.

The input to HI-RISE is a three dimensional grid, as illustrated in Figure 3. Spatial constraints, such as the location of vertical service shafts or internal spaces, are specified in terms of their location on the input grid. The intended occupancy of the building, and the wind and live load are the additional input information required by HI-RISE. Once the input has been specified, the interaction between the user and HI-RISE is graphical. The user participates in the selection of a structural alternative from the set of feasible alternatives generated by HI-RISE. The user may also request information about the building components of any feasible alternative.

HI-RISE is a developmental prototype ES that serves as a starting point for exploring the use of ES techniques for the preliminary structural design process. Currently, HI-RISE is being extended and implemented in Knowledge Craft on a Micro Vax II (Maher, 1984, 1986).

2.4.2. DESIGN SYSTEM FOR LOW-RISE INDUSTRIAL BUILDINGS: LOW-RISE

LOW-RISE aids in structural planning, preliminary design, and evaluation of industrial-type buildings. Planning consists of determining the components of the gravity and lateral load systems of various framing layouts that satisfy user input spatial constraints. Each alternative is ranked heuristically for comparison with other alternatives.

It was implemented in a combination of OPS5, LISP, and C. Heuristic knowledge, generation of framing schemes, and layouts for components of the gravity and lateral load systems were written in OPS5. More algorithmic parts such as analysis were coded in LISP. C was used to communicate with the database management system.

LOW-RISE relaxes the rigid spatial constraints of HI-RISE. The building is described in terms of large areas called departments, with each department identified by a column placement constraint. It first selects feasible structural configurations satisfying the column placement constraint separately for each department; it then attempts some global ''smoothing'' strategies to align the grid across departments. Finally, preliminary analysis, component sizing, costing, evaluation, and ranking are performed on each alternative. This is an operational prototype ES that has been developed with expertise supplied by experts from the Carnegie-Mellon University Architecture Department, American Bridge Company, and other industries (Camacho, 1985).

2.4.3. PRELIMINARY DESIGN OF FRAMEWORKS BY EXPERT SYSTEM (Ovunc, 1988)

A knowledge-based expert system (KBES) is used in the preprocessor of a general purpose software. The first part includes information related to the geometry, quality of the materials, and loads acting on the framework as data, whereas the second part contains the approximate sizes of all the members of the framework that are evaluated from the data provided in the first part. The second part, which constitutes the KBES, determines the member sizes using either the code requirements or certain approximate expressions. A cost analysis is also included in the second part, depending on the type of structures and the quality of materials used.

The software for the preliminary design is developed mainly in FORTRAN language in order to provide the ability to handle complex mathematics and to facilitate interfacing the various final design or other softwares. The modules related to the graphics are written in BASIC language.

2.4.3.1. Data Preparation

The first external data required by the preprocessor are related to (1) the selection of the computer type, (2) the processor to be interfaced, (3) the type of structural system to be analyzed, and (4) the type of analysis to be performed with or without the preliminary design. The remaining external data of the specific structural system under consideration include (1) the locations of the columns, (2) the types and qualities of the materials, (3) the dead loads, such as floor covering, floor finishing, etc., (4) the gravitational live loads, and (5) soil conditions, types of foundations, etc.

2.4.3.2. Preliminary Design

The preliminary design begins by checking the locations or spacings of the columns by considering the inference mechanisms or database, depending on the structural plans, number of floors, floor heights, externally applied loads, type and quality of materials used, etc. The thickness of the slabs is first evaluated for an optimum spacing of columns. The final design of the slab is performed by using the theory of plates, finite element method, code requirements, or the database. After the final design of all the slabs, the transfer of the gravitational loads from the slabs to the beams is evaluated. Besides the dead and live loads transferred from the slabs, the wall loads, self-weight of beams, and horizontal loads due to wind and earthquake are computed and the absolute sizes of the members estimated using moment coefficients for continuous beams under gravitational loads, the portal method for the frames under horizontal loads, inference mechanisms, or the database.

Figure 4 represents the variation of the moments of inertia of the beams on the abscissa with respect to the level n-i of the floors on the ordinate of the graph, where n is the floor number of the roof. The minimum beam moment of inertia appears on the roof floor, since the magnitudes of the gravitational loads on the roof are smaller than those on the lower floors. The sudden increase in the beam moments of inertia from the roof to the floor immediately below the roof is due to the increase in the rigidity of this floor due to the columns above the floor level and the increase in the gravitational loads from the roof to the lower floors. The axial forces in the columns increase from floor to floor in proportion to the tributary load area of the floor for that column. Figure 5 shows the variation of column moments of inertia at different floor levels. The column moments of inertia may remain constant for the very few top floors because of the minimum size requirements. The column sizes in the lower floors increase due to the increase in the gravitational loads and the effect of wind and earthquake. The variation of the column moments of inertia is different for the interior and exterior columns, since the axial forces in interior columns are larger than those in the exterior columns.

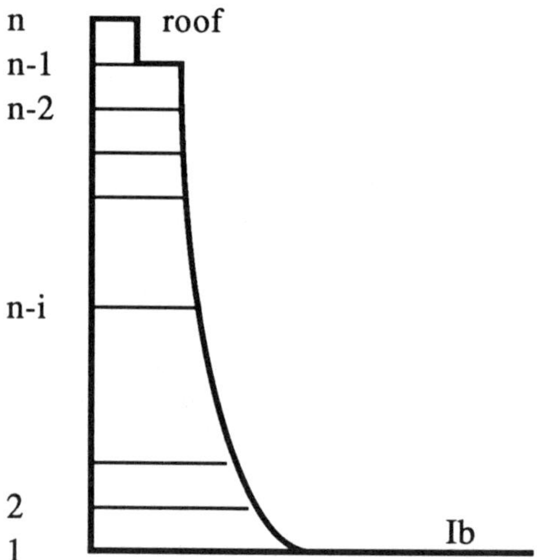

FIGURE 4. Variation of the beam moments of inertia. (From Ovunc, B. A., *Computing in Civil Engineering,* ASCE, New York, 1989, 28. With permission.)

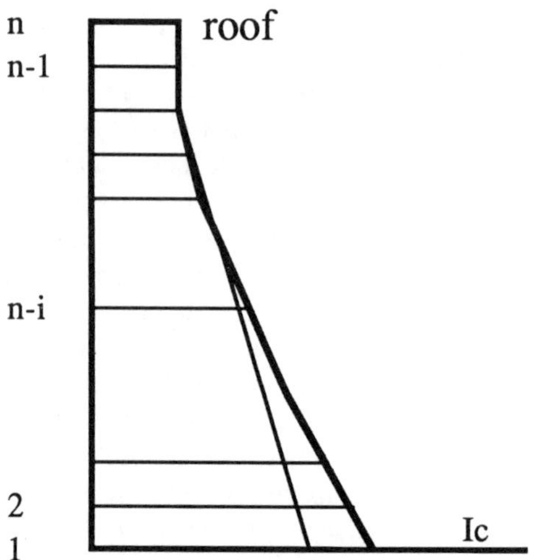

FIGURE 5. Variation of column moments of inertia. (From Ovunc, B. A., *Computing in Civil Engineering,* ASCE, New York, 1989, 28. With permission.)

The KBES is incorporated into the preprocessor, which is in a modular form that can be interfaced with the general purpose structural softwares.

2.4.4. BRIDGE DESIGN SYSTEM: BDES

The design of highway bridges is an ill-structured problem in which a large number of solutions are possible. Design decisions include selection of span type (continuous or simple), girder type (rolled beam, prestressed concrete, or plate girder), clearance, material types, etc. The expert system BDES (Bridge DEsign System) was developed to aid engineers in the decision, modeling, and analysis process of highway bridges in North Carolina. It incorporates expert knowledge to aid the decision process as well as knowledge of the serviceability and safety criteria of AASHTO (1983) and the state of North Carolina. The input to the system consists of graphical definitions of bridge geometry, bridge function, and the environment in which the bridge is to be constructed. Feasible alternatives to the problem are generated by the ES using approximations and assumptions. The designs are checked using the load factor approach, and decisions on the best design to be adopted are based on least weight. The system is capable of designing bridge superstructures of short to medium, simple, or continuous spans.

BDES was developed in PASCAL and uses a forward-chaining production rule approach, since it facilitates the decision-making process of design. Graphics are used for both the input process and output. The rule base is comprised of IF-THEN rules containing the information of experts as well as AASHTO bridge specifications and local ordinances of the state. The factual knowledge includes AASHTO requirements, material properties, and typical superstructure designs, whereas the heuristic knowledge includes rules for superstructure selection, girder spacing determination, and selection between simple or continuous span design. BDES is capable of selecting and proportioning short- to medium-span bridge superstructures (Welch and Biswas, 1986).

2.4.5. EXPERT SYSTEM FOR THE OPTIMUM DESIGN OF BRIDGE TRUSSES: BTEXPERT

BTEXPERT (Bridge Truss EXPERT) has been developed for the optimum design of four types of bridge trusses: Pratt, Parker, parallel-chord K truss, and curved-chord K truss for a span range of 100 to 500 ft. The system was developed using the Expert System Development Environment (ESDE) and the Expert System Consultation Environment (ESCE). The two programs, collectively referred to as the Expert System Environment (ESE), are a pair of complementary programs developed recently by the IBM Corporation. The first program is used to develop ES and, in particular, knowledge bases, whereas the second program provides the facilities for interactive execution of the ES. A graphics interface has been developed using the Graphical Data Display Manager (GDDM) (IBM, 1984). It was developed by interfacing an interactive truss optimization program developed in FORTRAN 77 to an ES environment developed in PASCAL/VS. Design constraints and the moving loads acting on the bridge are based on the American Association of State Highway and Transportation Officials specifications (AASHTO, 1983). The structure and functions of various components of BTEXPERT are presented in Figure 6.

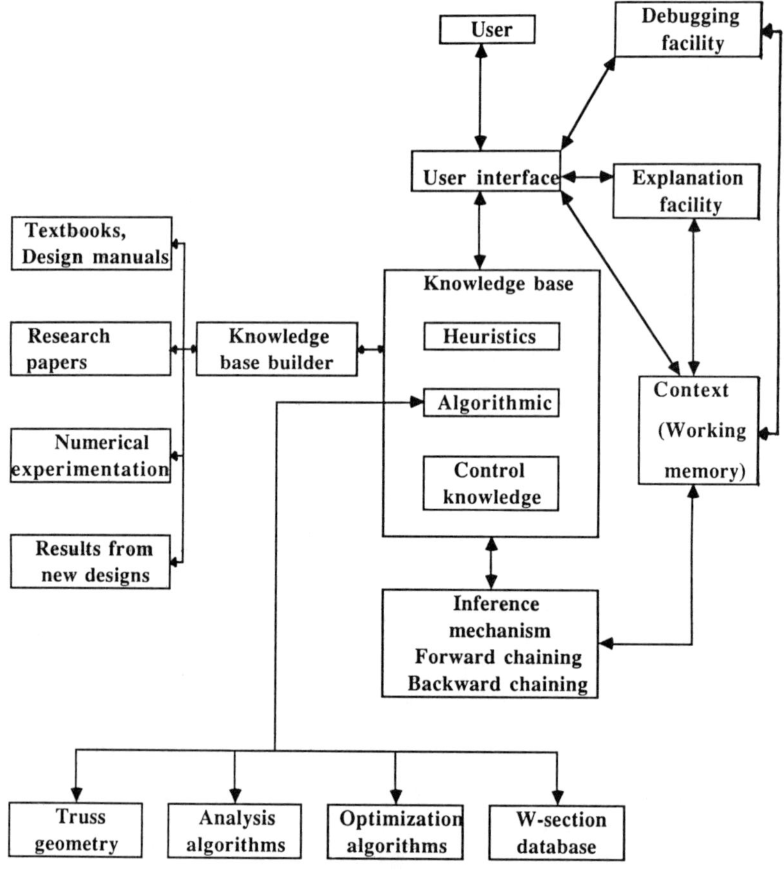

FIGURE 6. Architecture of BTEXPERT. (From Adeli, H., *Expert Systems in Construction and Structural Engineering,* Chapman and Hall, New York, 1988. With permission.)

2.4.5.1. Knowledge Base

The knowledge base of BTEXPERT consists of domain-specific knowledge and control knowledge. The domain-specific knowledge consists of rules and algorithmic procedures. The control knowledge consists of control commands for solving a problem. The rules consist of an IF part and a THEN part, or premise-action parts. Each rule represents an independent piece of knowledge. Knowledge representation consists of facts or parameters, rules, and focus control blocks (FCBs). FCBs are the main building blocks in the ESE.

Rules are classified into three categories:

Inference rules—The default type of any rule is the inference rule. These rules are processed either by forward or backward chaining.

Single-fire monitors—Single-fire monitors function independently without any reference to inference rules. The single-fire monitor is processed once a parameter in the IF part of a rule gets a value.

Multiple-fire monitors—These are processed exactly like single-fire monitors except that they may be executed many times.

2.4.5.2. Inference Mechanism

The ESE has both backward- and forward-chaining mechanisms for problem solving. In backward chaining, the facts for which values have to be determined are regarded as goals or subgoals. The goals and subgoals of an FCB are selected by the knowledge base builder. The rules are processed one at a time until all the goals and subgoals are found.

In the forward-chaining inference mechanism, the applicable inference rules are collected in a rule list. Known facts in the FCB are collected in a fact list. The ES processes the rule list in a top-down manner. Based on the values of the facts in the fact list, the THEN part is executed for rules having their IF parts satisfied. The fact list is subsequently updated. Processing of the rule list stops after one complete cycle through the applicable rule list if a single-cycle strategy is used; with the multiple-cycle strategy, the rules are processed in the applicable rule list again and again until the applicable rule list is empty or no remaining rules can be fired.

2.4.5.3. User Interface

The user interface is provided in the form of visual edit screens and menus in which the user has to type in the values of the required parameters at appropriate fields. The user can have graphical displays of the truss configuration with joint or member numbering (Figure 7), influence line diagrams (ILDs) for various member axial forces and joint displacements, and the design AASHTO live loads.

2.4.5.4. Explanation Facility

The explanation facility helps the user to examine the reasoning process. The explanation consists of both the RULE text and RULE comments coded by the knowledge base builder. The explanation facility commands are:

EXHIBIT—It displays the current value(s) of a specific parameter.

HOW—It displays an explanation of how the system determined a value for a parameter. Figure 8 shows an example of the explanation generated by BTEXPERT in response to the HOW command during a sample consultation.

WHY—It displays an explanation of why the system is asking a given question (Figure 9).

WHAT—It displays more information about a given parameter (Figure 10).

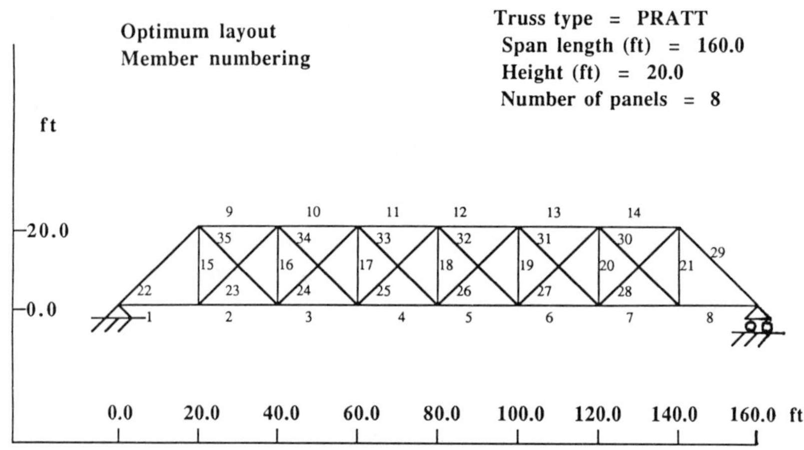

FIGURE 7. A sample Pratt truss plotted by BTEXPERT. (From Adeli, H., *Expert Systems in Construction and Structural Engineering,* Chapman and Hall, New York, 1988. With permission.)

```
                    Focus : FCB11 (1)

              .... How ....
I assigned value to allowable stress range in fatigue of
FCB11 by
1. Rule RULE0039 which states that
If AASHTO LIVE LOAD = 'HS-15'
 or AASHTO LIVE LOAD = 'HS-20'
Then Mumber of stress cycles = 500000
and Allowable stress range in fatigue = 24.
This rule is based on the AASHTO specification.
As a result of this rule
Allowable stress range in fatigue assigned = 24 (1)
                    To continue Consultation, Press ENTER

  ==>

PF1  Help
PF2  Review
PF3  End
PF4  What
PF5  Question
PF6  Unknown
PF7  Up
PF8  Down
PF9  Tab
PF10 How
PF11 Why
PF12 Command
```

FIGURE 8. Example of explanation generated by BTEXPERT in response to HOW it arrived at the value of the parameter allowable_ stress_ range_ in_ fatigue. (From Adeli, H., *Expert Systems in Construction and Structural Engineering,* Chapman and Hall, New York, 1988. With permission.)

2.4.5.5. Debugging Facility

The ESDE knowledge acquisition editors check each parameter, rule and FCB for syntax errors whenever they are typed into the system. However, the knowledge base builder should ensure that the knowledge base is consistent and complete, since inconsistencies between individual rules or in various parts of a rule are not checked by the ESE. The knowledge base builder can use the TRACE facility to debug errors detected in the results.

Focus : FCB2 (1)

Enter the 'Location of bridge'

(Choose one of the following:)

.... State-Road
.... Trunk-highway
.... Interstate

==> WHY LOCATION OF BRIDGE

To continue Consultation, press ENTER

PF1 Help
PF2 Review
PF3 End
PF4 What
PF5 Question
PF6 Unknown
PF7 Up
PF8 Down
PF9 Tab
PF10 How
PF11 Why
PF12 Command

Focus : FCB2 (1)

....WHY....
I am asking about BRIDGE LOCATION of FCB2 to find AASHTO LIVE LOAD which I am trying to determine.

These rules are used for this line of reasoning.

..RULE RULE0012..
If BRIDGE LOCATION is 'Trunk-highway'
Then AASHTO LIVE LOAD = 'HS-15'.

(Choose one of the following:)

.... State-Road
.... Interstate
.... Trunk-highway

PF1 Help
PF2 Review
PF3 End
PF4 What
PF5 Question
PF6 Unknown
PF7 Up
PF8 Down
PF9 Tab
PF10 How
PF11 Why
PF12 Command

FIGURE 9. Example of explanation generated by BTEXPERT in response to WHY it is asking the value of the string parameter bridge_ location. (From Adeli, H., *Expert Systems in Construction and Structural Engineering,* Chapman and Hall, New York, 1988. With permission.)

Focus : FCB3 (1)

....What....
If the bridge location is in an area of high corrosion, the recommended choice of steel will be M244. It should be noted that the relative costs of M183, M223, M222,and M244 are 1.0, 1.15, 1.33, and 1.73, respectively and the resistance to corrosion is poor, not bad, good, and best, reapectively (Ref.Heins 1979).

(Choose one of the following:)

_x_ M183
___ M223
___ M222
___ M244

==>

FIGURE 10. Example of WHAT explanation command for providing additional information about the parameter steel_ type. (From Adeli, H., *Expert Systems in Construction and Structural Engineering,* Chapman and Hall, New York, 1988. With permission.)

2.4.5.6. Knowledge Acquisition

Domain knowledge is partly obtained from textbooks, design manuals, design specifications (e.g., AASHTO, 1983), research papers, and journal articles. Besides these sources, gaps in the knowledge base are filled with optimum values of primary design parameters obtained from detailed nu-

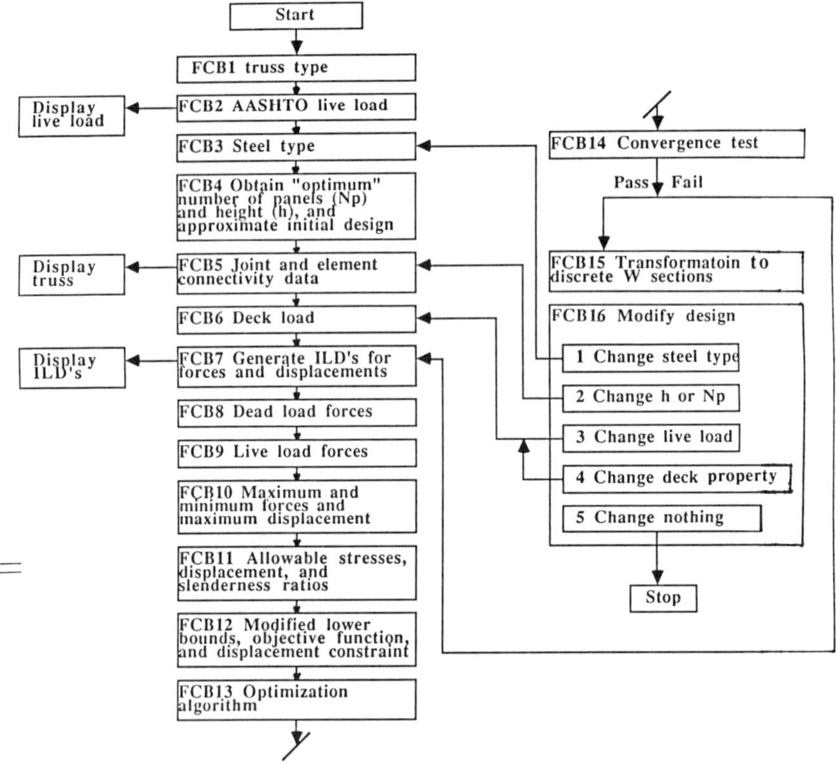

FIGURE 11. Structure of FCBs. (From Adeli, H., *Expert Systems in Construction and Structural Engineering*, Chapman and Hall, New York, 1988. With permission.)

merical machine experimentation in the problem domain. The machine experimentation can be conducted using the software program IOTRUSS developed in FORTRAN 77 for layout optimization of trusses. The optimum values for the height, number of panels, and initial cross-sectional areas of truss members for various span lengths, AASHTO live loads, and grades of steel are subsequently used in the knowledge base of BTEXPERT.

2.4.5.7. Knowledge Base Development

The rules and procedures used in BTEXPERT are classified into a number of FCBs (Figure 11). Each FCB contains rules and procedures for a specific task. FCBs are used to classify all the rules and procedures required in an ES according to their intended uses and sequences of application. For example, the rules for selecting the right type of truss for the span length specified by the user are:

If Span_ length > = 100 and Span_ length < = 200
Then Recommended_ Truss_ type is 'Pratt'

If Span_ length > 300 and Span_ length < = 380
Then Recommended_ Truss_ type is 'Parallel-chord K truss'

Sample rules used in FCB2 for selecting the right type of design live loads for the bridge under consideration are:

If Bridge_ location is 'State-Road' and Traffic is 'Light'
Then AASHTO_ live_ load is 'H-15'

If Bridge_ location is 'Interstate-Highway'
Then AASHTO_ live_ load is 'HS-20'

More rules on FCBs are given in Adeli and Balasubramanyam (1988).

2.4.5.8. Mathematical Optimization

The optimum design of a bridge truss consists of selecting the right combination of the cross-sectional areas of the truss members so as to satisfy all the design constraints and produce a least-weight truss. The allowable compressive stresses and the slenderness limitations provided by AASHTO specifications involve the minimum radius of gyration of the cross section. Using these optimum cross-sectional areas obtained from BTEXPERT and heuristic rules, wide flange sections are selected for truss members from a database containing the W-Sections given in the American Institute of Steel Construction manual (AISC, 1978).

BTEXPERT is currently being extended to the optimum overall design of steel truss and plate girder bridges. Heuristic rules and procedures are being developed to improve the efficiency and accuracy of the optimization process, and for classification of constraints into inactive, partially active, active, and violated constraints (Adeli, 1988).

2.4.6. RETAINING WALL DESIGN: RETWALL

2.4.6.1. Introduction

The RETWALL expert system (ES) was developed to provide expertise in the specific area of retaining wall structures. Its capabilities include consulting on the choice of retaining structures for a given set of user inputs and performing the preliminary design. The choices of retaining structures in RETWALL are brick, blockwork, gabion, gravity, reinforced earth, reinforced concrete, and sheet pile.

The ES control lies in the existing ES shell BUILD, developed by the Department of Architectural Science, University of Sydney, Australia; it uses a backward-chaining production rule system written in Quintus Prolog. The system employs graphics procedures, written in C, to display preliminary designs as well as displays to enhance the input process.

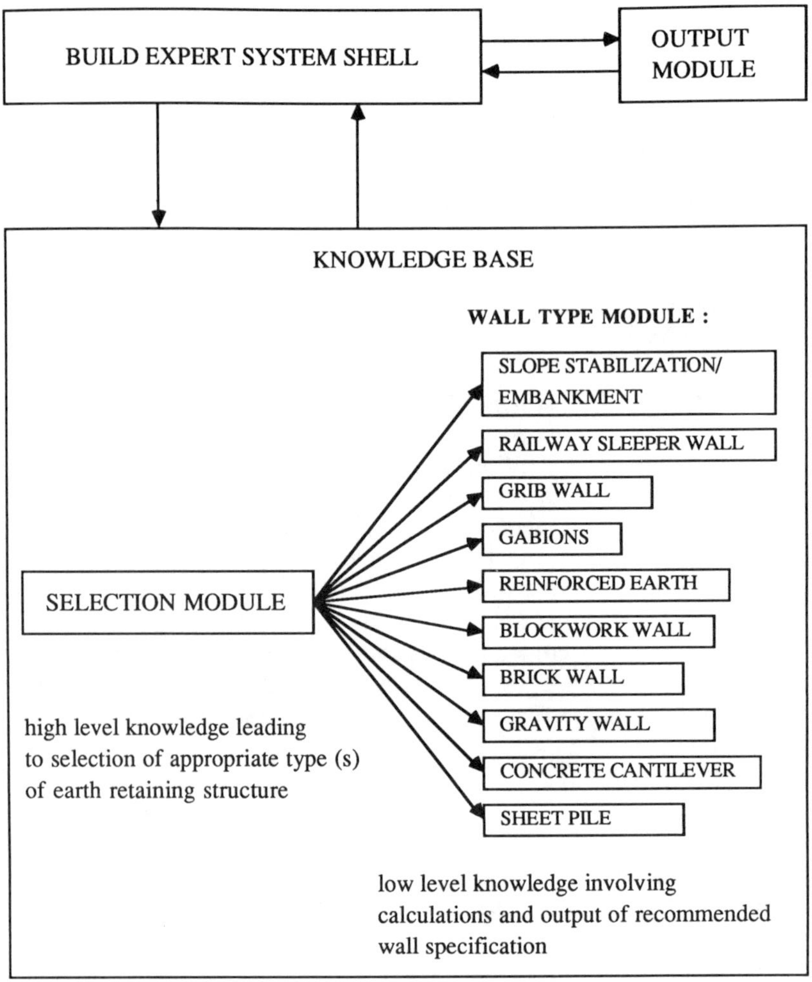

FIGURE 12. Outline of an ES for the selection of earth-retaining structures. (From Hutchinson, P., An Expert System for the Selection of Earth Retaining Structures, M.S. thesis, University of Sydney, Department of Architectural Science, Australia, 1985. With permission.)

The system consists of two main modules, the high end and the low end. The primary function of the high end of the system, which contains the rule base and inference mechanism, is to select the particular retaining structure to be used. The lower-end module consists of the routines that perform the preliminary design of the different retaining wall options. Presently, the lower-end routine has the capability to design only blockwork walls. Design in the lower-end routines is performed using design tables in the knowledge base of that module. The major limitation of the system is the lack of an evaluation of design alternatives (Gero, 1986; Hutchinson, 1985). Figure 12 shows the overall concepts of the system.

2.4.6.2. The Selection Module

The selection module contains the higher-level knowledge obtained from the literature review and interviews of experts, which is concerned with the selection of the various types of earth-retaining structures. Its rules are formulated such that they permit firing of the lower-level blockwork module only when it has been determined that a blockwork wall is suitable for the given application. Currently, if a type of structure other than blockwork wall is identified as suitable, a message is output to that effect and no further investigation of that type is conducted because the relevant lower-level modules have not been written.

The rules in the selection module can be divided into a number of blocks that provide knowledge on:

1. Typical site conditions and geometric parameters of the site for the various applications where an earth-retaining structure may be required
2. Whether an earth-retaining structure is required
3. The types of structures that should be investigated for a given application
4. Each type of structure considered and the factors that affect the selection of that type
5. Various other considerations that affect wall selection such as terracing, surcharge loading, and soil properties

A schematic layout of all the knowledge blocks in the higher module is given in Figure 13.

Knowledge on all the individual types of structures (brick wall, blockwork wall, crib wall, gabions, gravity wall, railway sleeper wall, reinforced earth, reinforced concrete wall, and sheet piling) is included in the system, although the amount of knowledge on each structure type reflects the amount of knowledge available from both the literature and the human experts. Hence, there is more knowledge in the rules on reinforced earth, which is rapidly gaining popularity, than in the rules on gravity walls, which are hardly used now.

The knowledge on typical site conditions is not only provided in the rules of the system, but is also displayed by simple drawings produced by a C-language procedure and called from within the ES at the appropriate time. Three different drawings, depending on the application given by the user, can be produced by the procedure, each showing a number of possible alternative site conditions. The user is then asked to indicate the site case most applicable and provide the physical dimension data shown on the diagram.

Figure 14 shows the flowchart for the knowledge that determines whether an earth-retaining structure is required. One of the main points to emerge from the interviews of experts was that an earth-retaining structure should only be employed if an embankment or cut cannot be used, or if there is some general reason for not wanting an embankment or cut. The knowledge block on whether an earth-retaining structure is required attempts to establish if an

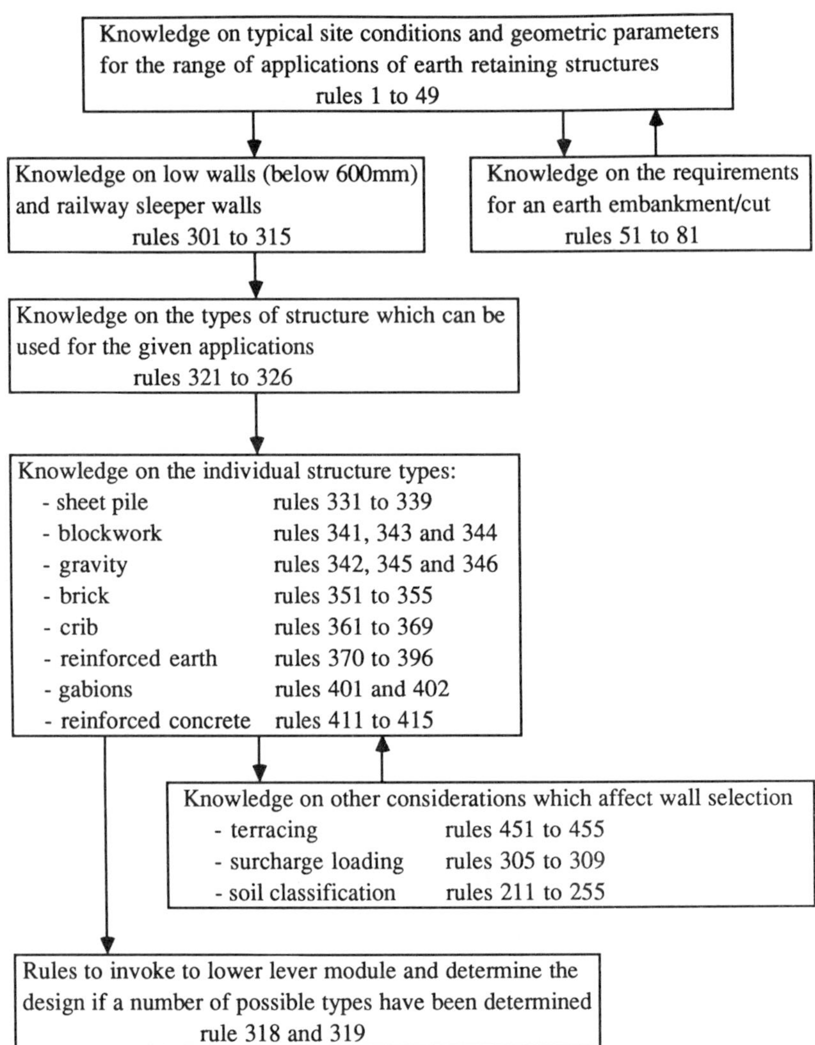

FIGURE 13. Schematic layout of all the knowledge blocks in the higher-level module. (From Hutchinson, P., An Expert System for the Selection of Earth Retaining Structures, M.S. thesis, University of Sydney, Department of Architectural Science, Australia, 1985. With permission.)

embankment or cut could be constructed. If not, then it is determined by default that an earth-retaining structure is required.

The knowledge on the types of structure suitable for a given wall application provides a higher-level control on the search and determines the order in which the various wall types are considered, and which types are considered for every application. If the types considered by these rules prove to be

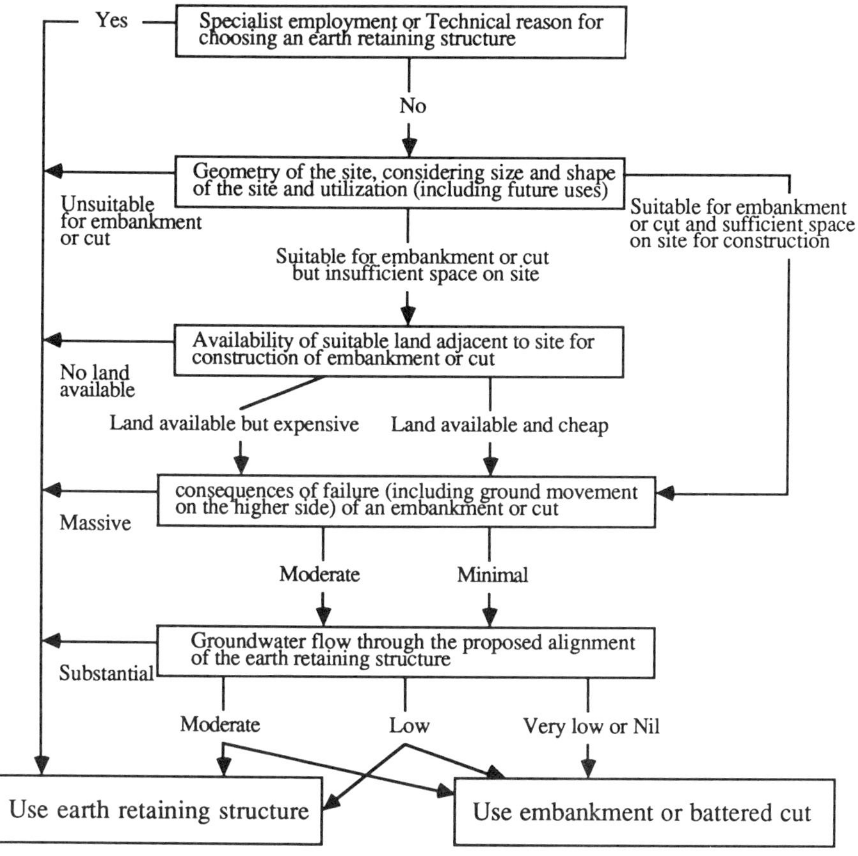

NOTE: Where there are two possible courses indicated, the course chosen will depend on the results of earlier courses taken. Generally if an earlier decision involved a course on the left of the center of the page, further decisions on the left of the page will result in an earth retaining structure being employed.

FIGURE 14. Flowchart for knowledge on the requirement for earth-retaining structures. (From Hutchinson, P., An Expert System for the Selection of Earth Retaining Structures, M.S. thesis, University of Sydney, Department of Architectural Science, Australia, 1985. With permission.)

infeasible, then the system will determine that the design is beyond its knowledge and stop execution of all the other possible, but not feasible, rules for evaluating a design.

The knowledge used in this block is formulated as rules such as:

```
r325 (if
    'earth retaining structure' is_ required and
    'type of application for wall' is_ A and
```

```
  'type of application for wall' is_ marine and
  evaluated)
     'Sheet pile is suitable for this application' and
     'Reinforced concrete wall is suitable for this application' and
     'Reinforced earth is suitable for this application') and
 not ('Sheet pile is suitable for this application') and
 not ('Reinforced concrete wall is suitable for this application') and
 not ('Reinforced earth is suitable for this application')
 then
 design of earth retaining structure' is_
       'beyond knowledge of this system').
```

The rules on the individual types of structure vary with the amount of knowledge obtained on the structures, but generally include a range of heights applicable for the structure, the types of application for which the structure may be used, the aesthetic suitability of the structure, and the availability of labor and materials for the structure. A typical example is

```
r351 (if
 'height of earth retaining structure (in mm)' is_ less_ or_ equal_ to_1500
 'Brick wall is aesthetically acceptable' and
 'Labor and materials are available for brick wall'
 then
 possible ('type of earth retaining structure type is_ 'brick wall') and
 'Brick wall is suitable for this application').
```

The final block of rules provides knowledge on such things as terracing, surcharge loading, scale of the project, and soil conditions, which can then be used by the other rules. Some of these rules may not be required in the case of an experienced user who may give the answers they provide directly. Generally, they are employed by the user asking "how" to the relevant question in one of the selection rules.

Examples of the rules for terracing and related rules are

```
r369 (if
    'type of application for wall' is_ A and
    'type of application for wall' is_ domestic or commercial or
    industrial or road or railway and
   'height of earth retaining structure (in mm)' is_ greater_ than
   15000 and
   not ('Reinforced earth is suitable for this application') and
   'slope ratio' is_ greater_ or_ equal_ to 1.83 and
   not ('The number of terraces required, considering aesthetics
   and space' is_ nil or 1 or 2), and
```

TABLE 2
Example of the Design Charts Used for Blockwork Walls

Backfill type	Height (m)	Wall type	Footing width (mm)	Wall dimensions	
				V-bars	X-bars
1	1.0	150	750	S16 @ 400	S16 @ 400
	1.4	200	900	S16 @ 400	S16 @ 400
	1.8	200	1050	S16 @ 400	S16 @ 400
	2.2	200	1300	S16 @ 400	S16 @ 400
	2.6	and	1550	S16 @ 400	S16 @ 400
	3.0	300	1750	S20 @ 400	S20 @ 400
	3.2	300	1850	S20 @ 200	S20 @ 400
3	1.0	150	1000	S16 @ 400	S16 @ 400
	1.4	200	1150	S16 @ 400	S16 @ 400
	1.8	200	1400	S16 @ 400	S16 @ 400
	2.2	200	1600	S16 @ 400	S16 @ 400
	2.6	and	1850	S20 @ 400	S20 @ 400
	3.0	300	2000	S24 @ 200	S24 @ 400
4	1.0	200	1150	S16 @ 400	S16 @ 400
	1.4	200	1450	S16 @ 400	S20 @ 400
	1.8	and	1750	S20 @ 400	S20 @ 400
	2.2	300	2300	S24 @ 200	S24 @ 400

Note: This chart applies for a base type 1 wall (see Figure 15).

'Crib wall is aesthetically acceptable' and
'Labor and materials are available for crib wall'
then
possible ('type of earth retaining structure' is_ 'crib wall'
and
'crib wall is suitable for this application').

r453 (if
'slope ratio' is_ greater_ or_ equal_ to 1.33 and
'slope ratio' is_ less_ than 1.5
then
'maximum number of terraces allowed' is_ 2).

2.4.6.3. The Blockwork Module

The blockwork module uses knowledge contained in design charts to produce preliminary designs for reinforced concrete masonry-retaining walls from 1.0 to a maximum, depending on the backfill soil used, of 3.2 m in height. A feature of this module is the output produced, which not only gives wall parameters, but also a scaled, dimensioned drawing showing reinforcing bar requirements.

The design charts used to produce the majority of the rules in this module give footing width, reinforcing bar requirements, and wall thickness requirements for a given wall height, footing type, and backfill soil type. An example of one of the charts used is shown in Table 2.

The blockwork module contains knowledge to

1. Classify the backfill into soil types given by Terzahgi and Peck (1967)
2. Check that the allowable subgrade bearing pressure is not exceeded
3. Select the most appropriate wall footing type for the given site conditions
4. Select the appropriate reinforced concrete masonry (blockwork) wall design parameters for the given conditions

The effects of backfill soil in exerting pressure on the retaining wall are based on empirical charts for active soil pressure given by Terzahgi and Peck for walls less than 6 m in height. The gradings range from granular soil with little or no fines (backfill type 1) to medium or stiff clay deposited in chunks and protected from water penetration (backfill type 5). The lower the type, the more suitable it is for use as backfill, and the smaller will be the section of wall required to retain it due to the lower active soil pressures produced.

The system uses either verbal descriptions of the backfill soil or the Unified Soil Classification of the soil to grade the backfill as type 1 to 5. For example, a "backfill type" is "sand or gravel containing some silt" or Unified Soil Classification GP-GM, GW-GM, SW-SM, or SP-SM. To obtain the Unified Soil Classification, a module of about 40 rules (adapted from Burnham et al., 1984) has been included that gives the classification based on the results of sieve analysis and laboratory tests.

Examples of the rules for backfill type are

```
r.261 (if
    'backfill to be used' is 'sand or gravel with little or no fines'
    then
    'backfill type' is_ 1).

r262 (if
    'backfill to be used' is_ other and
    'soil classification of backfill' is X and
    'soil classification of backfill is_ 'GW' or 'GP' or 'SW' or 'SP'
    then
    'backfill type' is_ 1).
```

The first of these two rules is self-explanatory. When this rule is "fired", the user will be asked what the backfill to be used is and given the five options for the five soil types along with the option of answering "other" and having the system determine the Unified Soil Classification. If the user answers "other", the first rule fails and the second rule is invoked. The first line of this rule will succeed and the system will then attempt to determine the soil classification. The two lines on "soil classification of backfill" are required for the same reason discussed for the "type of application for wall" in Section 2.4.6.2—to ensure evaluation of this predicate by the BUILD expert system (ES) shell.

The allowable subgrade bearing pressure for the walls given by the design charts used must not be below 125 kPa. To ensure that this restriction is compiled with the rules dealing with footing-type selection, the subgrade allowable bearing pressure is first determined. If the user cannot provide a direct answer in kilo pascals, rules giving approximate allowable bearing pressures based on charts given by Carter (1983) will be invoked which match verbal descriptions of the subgrade soil with a minimum approximate bearing pressure.

These rules are self-explanatory and take the form:

```
r105(if
    'soil beneath wall footing' is_ 'firm clay'
    then
    'subgrade allowable bearing pressure (kPa)' is_ 130).
```

A note is included with the display of the question on the "soil beneath wall footing" to give some rules of thumb for estimating the bearing pressure and matching the verbal description.

Four different wall-footing types, as shown as Figure 15, are considered by the blockwork module. The most economical and preferred one is type 1, while type 4 is preferred if only limited space is available for excavation and construction behind the face of the wall. Type 2 and 3 wall footings are applied in boundary wall situations where all the available space on a site is required and the wall footing cannot pass beneath some boundary or site restriction. The knowledge of site geometry and restrictions required by the rules that determine the wall footing type (base type) is obtained by the selection module and is thus already in the facts base of the ES. These rules take the form:

```
r271 (if
    'subgrade allowable bearing pressure (kPa)' is_ greater_ than 125 and
    'site case most applicable (as shown in the diagram)' is_ 1 and
    'horizontal distance shown (d) (in mm)' is_ greater_ or_ equal_ to 500
    then
    'base type' is_ 1).
```

The "subgrade allowable bearing pressure (kPa)" has already been discussed; these rules ensure that it is instantiated and checked before the design for a blockwork wall can be produced. The "site case" and "horizontal distance" refer to a drawing produced by the selection module that the user would already have answered questions on by the time this rule is "fired". Hence, the user would only have to provide the subgrade allowable bearing pressure and the system would automatically deduce the base type.

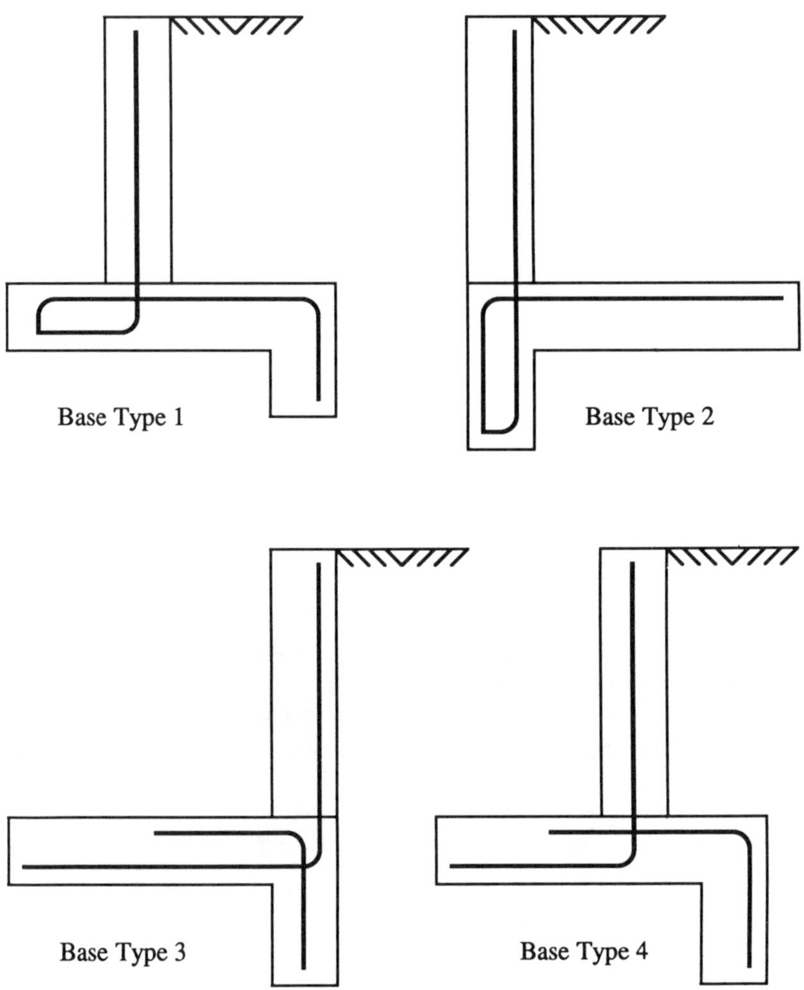

FIGURE 15. The different wall footing (base) types used. (From Hutchinson, P., An Expert System for the Selection of Earth Retaining Structures, M.S. thesis, University of Sydney, Department of Architectural Science, Australia, 1985. With permission.)

The final block of rules in the blockwork module form the major part of the module providing design parameters for the wall and invoking the C-language graphics procedure to produce a scaled, dimensioned drawing showing reinforcing bar requirements. A typical example of these rules is

r136 (if
'height of earth retaining structure (in mm)' is_ greater_ than 2600 and
'height of earth retaining structure (in mm)' is _less _than _or _equal _to

3000 and
'base type' is_ 1 and
'backfill type' is_ 3
then
'blockwork wall type' is_ 300 and
'footing width' is_ 2000 and
'V-bars' is_ 'S24 at 200' and
'X-bars' is_ 'S24 at 400' and
draw).

The blockwork module is invoked by the selection module trying to prove that the "'blockwork wall type' is_" X. In other words, the selection module wants to find a value for the blockwork wall type, and that value will be instantiated by the first of the rules of the type shown above which succeeds. In proving the blockwork wall type, all of the other predicates in the consequent part of the above rule will also be instantiated and the six design parameters required to describe a blockwork wall will thus be known. These parameters are the height, base type, blockwork wall type, footing width, V-bar, and X-bar requirements.

The final predicate in the above rule, "draw", is recognized by the BUILD ES shell and a Prolog rule in the shell is "fired" to call the C-language graphics procedure, converting the Prolog form of each of the design parameters into C-arguments for the procedure.

Having succeeded in proving that the "'blockwork wall type' is_" X, control of the ES returns to the selection module.

2.5. CONCLUDING REMARKS AND SUGGESTIONS FOR FURTHER WORK

ES applications to structural systems are research oriented rather than commercial oriented, and concerned with the representation of design knowledge and the design process. The sample systems presented here are applications to the structural design of buildings, retaining-wall design, bridge design, and design of frameworks. The potential use of ES for structural design depends on the complexity of the design problem. The ES approach will aid in the selection process of design problems in which the number of alternative solutions is small.

Knowledge-based expert systems (KBES) deal only with *shallow* knowledge, i.e., empirical associations. KBES environments could be more closely coupled with the algorithmic programs that would contribute the deep, causal knowledge. KBES has the potential to be used not as standalone programs, but as intelligent pre- and postprocessors for existing programs such as finite element analyzers. The KBES framework would provide increasing user interface, explanation, and knowledge acquisition.

Chapter 3

DESIGN STANDARDS PROCESSING

3.1. INTRODUCTION

Design standards play an important role in the design of engineering systems. A design configuration must be checked against all applicable standards to ensure that it is acceptable. Previous research on design standards has been conducted to improve the (1) representation and organization of standards, (2) analysis of standards, and (3) use of standards. Standards are often modeled using three tools: decision tables, information networks, and an organization system (Fenves, 1980; Harris and Wright, 1980; Rasdorf and Fenves, 1980).

3.2. GENERIC DESIGN STANDARDS PROCESSING

The processing of design standards in an expert systems (ES) environment was initially investigated by building two knowledge-based expert systems (KBES): (1) Query Monitor addresses the issue of semantics of data retrieval from engineering databases and (2) Roofload Checker performs design conformance checking utilizing a standard.

3.2.1. QUERY MONITOR

The American Institute of Steel Construction specification addresses a number of different types of stresses within a structural steel member, including tension, shear, compression, bending, and bearing (AISC, 1978). Depending upon the constraints on shape, cross section, loading, etc., any one of a number of equations can be used to determine the allowable stress for a specific structural steel member. A database problem arises when the engineer issues an F_b data retrieval request. Query Monitor was identified as a framework to combine a database with a set of design specification constraints that govern the retrieval of data from engineering databases (Rasdorf and Wang, 1986). Query Monitor architecture was developed using the M.1 ES building tool (Teknowledge, 1985). The knowledge representation consists of production rules and facts. The inference engine utilizes a goal-driven control strategy. Figure 1 shows a decision table from Provision 1.5.1.4 of the AISC specification. The first column of the table was recast in production rule format as follows:

If the axis about which a member is being bent is major *and*
the connection of the web and flange is continuous *and*
the width thickness ratio for exceptions is ok *and*

Axis = major	Y	Y	Y	Y	Y	N
ConOK	Y	N	Y	Y	Y	Y
Bf_Tf_RatioOK	Y	Y	N	Y	Y	Y
D_T_RatioOK	Y	Y	Y	N	Y	Y
LcOK	Y	Y	Y	Y	N	I

Fb = 0.66*Fy	X					X
Except_Ten		X	X	X	X	
Except_Comp		X	X	X	X	

FIGURE 1. Decision table for AISC specification provision 1.5.1.4. (From Rasdorf, W. J. and Wang, T. E., *J. Comput. Civ. Eng.*, 2(1), 68, 1988. With permission.)

the depth thickness ratio is ok *and*
the laterally unsupported length is ok
*Then_* the allowable bending stress = 0.66 F_y

A complete program listing as well as several sample execution logs are given in the *Query Monitor User's Guide* (Wang and Rasdorf, 1985).

3.2.2. ROOFLOAD CHECKER

The Roofload Checker was developed to study the performance of a production system based on a data-driven control strategy to check designs. It consists of two subprograms, Roof Checker and Roof Reporter. The engineer describes the roof design using datum-value pairs that are stored in the context. Roof Checker then checks the roof design by matching the input against the production rules converted from the Building Officials and Code Administrators International building code (BOCA, 1984) to determine whether the design conforms to the standards it incorporates. However, it does not provide any feedback after its operation. The result after design checking by Roof Checker is stored in an external file. After Roof Reporter is invoked, the data from the file are then reformatted and displayed on the monitor screen.

Roof Checker and Roof Reporter were written in the OPS5 knowledge engineering language, and the knowledge representation scheme consists of production rules. Either the data-driven or the goal-driven control strategy can be implemented in OPS5. As an example, the requirements of Table 910 of the code are directly cast in production rule format in the Roof Checker as follows:

*If_* the shape of the roof is pitched *and_*
$4 \leq$ the slope of the roof $<$ 12 in/ft *and_*

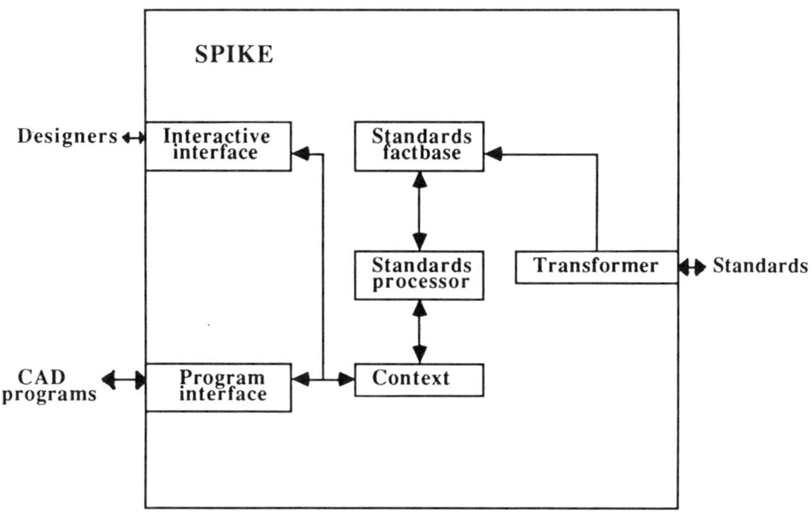

FIGURE 2. Architecture of SPIKE standards processing system. (From Rasdorf, W. J. and Wang, T. E., *J. Comput. Civ. Eng.*, 2(1), 68, 1988. With permission.)

$0 \leq$ the tributary loaded area for structural member < 200 ft^2 *and*
the designed roof load ≥ 16 psf
Then the roof is OK

More details of Roof Checker as well as Roof Reporter are reported by Wang (1986).

3.2.3. GENERIC STANDARDS PROCESSING IN A KNOWLEDGE-BASED EXPERT SYSTEM ENVIRONMENT: SPIKE

The architecture of SPIKE consists of two functions: (1) performing design conformance checking and (2) determining allowable value ranges for undetermined design datums. Figure 2 shows the typical components of SPIKE architecture. It uses provisional and organizational facts for its knowledge base. Because the knowledge base is implemented in the factual format, it is called the standards factbase of SPIKE. As in a typical expert system (ES), the standards factbase is used by an inference engine as it manipulates the context. The set of production rules encoded specifically for processing the generic standards factbase is referred to as a standards processor. The transformer, which is the knowledge acquisition facility in SPIKE, translates the knowledge from the decision table format of a standard to the internal representation of the factbase. The context is the short-term memory containing design-specific information entered by the interfaces (interactive and program) or generated by the inference engine. The interactive interface provides a command language to enable the designer to describe a design, or to query the system to obtain information about the design or the governing standards. The program interface provides a similar functionality for CAD programs.

SPIKE has been implemented as a research prototype using OPS5, whose operation is governed by pattern matching. The user enters, as input, sets of datum-value pairs describing the design under review. When the user indicates there is no additional input, SPIKE performs data generation, analysis, and design conformance checking, and the results are displayed on the screen. The user can then elect to quit or continue, revising the design by entering updated datums or new datum-value pairs, and the cycle can be repeated as many times as necessary until a design is derived that completely conforms to the governing standard. The detailed implementation of SPIKE is described by Rasdorf and Wang (1986, 1988).

3.3. AASHTO BRIDGE-RATING SYSTEM

An ES that carries out the rating of simply supported highway bridges with reinforced concrete decks and prestressed concrete I-beams is under development at Lehigh University. The effects of vehicular or overloaded vehicular traffic are taken into account. The expert knowledge stored in the database includes AASHTO bridge-rating provisions, extensive data on the overload of prestressed concrete highway bridges, and heuristics essential to decision-making strategies. The database is structured in a two-dimensional spreadsheet format. The basic approach involves a forward-chaining search of the database for a bridge rating (i.e., AASHTO, past case histories, and Grillage Analogy). At the exhaustion of the database, if the rating quality is unsatisfactory, finite element algorithms are triggered and the bridge is treated as a new design problem. The system is an operational type and is written in structured FORTRAN (Kostem, 1986).

3.4. AUSTRALIAN MODEL UNIFORM BUILDING CODE: AMUBC

Design codes contain a large amount of causal and experiential knowledge. Typically, the amount of information in a code is large and represents the best effort on the part of the writers to organize it in a clear fashion. Even with this effort, codes tend to be unstructured, complex, and difficult to interpret by many engineers. AISC, ACI, BOCA, etc. are examples of codes that could appear unstructured and are hard to follow. The ability to use a code to its full potential depends on the experience and expertise of the individual using it. The primary motivation for the development of design codes as ES is to produce computer systems that will aid not only the engineer and designer, but also the local authorities in administration of these codes.

The prototype ES representing only a part of the entire AMUBC is run on an ES shell written in Prolog 1 on an 8088/8086 microcomputer in an MS-DOS environment that needs a minimum of 128 kbytes. A production system approach has been used for knowledge-base development since this rule-based approach facilitates the modeling of the information as it is typically presented in building codes. The system is capable of both forward and backward chaining through the rule base. The domain-independent meta-knowledge, which is an important feature of the ES, provides the user with the capability of determining the scope of the information relevant to the problem and nature of the knowledge in the domain of the system. One disadvantage of the ES is its lack of interrupt capabilities for explanation facilities. Work is now in progress in determining better representations and expansion to include more of the AMUBC (Rosemann, 1985, 1986).

3.5. KNOWLEDGE-BASED STANDARDS PROCESSOR: SPEX

SPEX is a knowledge-based structural component design system that basically selects requirements, generates constraints, and then satisfies those constraints to find a set of values for the properties of the component. The system is knowledge based because designer expertise is used to select behavior limitations for detailed design in which the properties of _all_ structural components are determined subject to the satisfaction of structural integrity and functionality constraints.

It is implemented as a blackboard system because the blackboard architecture facilitates the integration of knowledge-based and algorithmic subprocesses in the component design process. The architecture of SPEX is shown in Figure 3. Task specification, design focus hypothesis, standard requirements, constraints, and solution form the five levels of abstraction in the blackboard. The knowledge base in SPEX is divided into the design process modules and design knowledge. The design focus module generates a design focus hypothesis using a set of expert rules. The requirement retrieval module generates (1) a list of requirements that must be checked and (2) a list of requirements that are translations of the behavior limitations within the design requirements. The constraint set generation module generates a set of constraints from the design requirements. The constraint set satisfaction module determines the optimal component design within the solution space defined by the constraint set using either a nonlinear constraint satisfaction routine, OPT (Biegler and Cuthrell, 1985), or a knowledge-based database interface, KADBASE (Howard, 1986). The conformance verification module checks the resulting component design not only for conformance with design requirements, but also for conformance with all applicable standard requirements. If violated requirements are found, a backtracking situation would occur and the design focus module should be invoked to alter its hypothesis such that the violated requirements become design requirements.

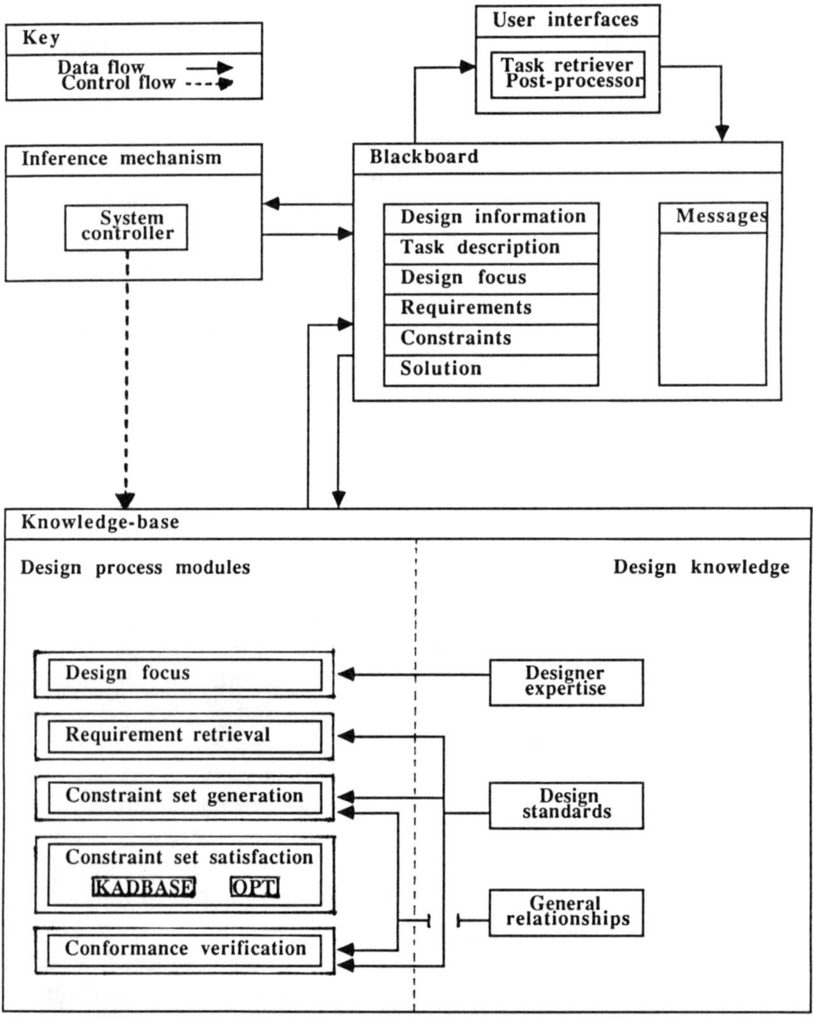

FIGURE 3. The functional modules of SPEX. (From Garrett, J. H., Jr., SPEX—A Knowledge-Based Standard Processor for Structural Component Design, Ph.D. thesis, Carnegie-Mellon University, Pittsburgh, PA, 1986. With permission.)

The design knowledge in the knowledge base consists of (1) designer expertise for the generation, completion, and modification of design focus hypotheses, (2) design standards, and (3) general relationships, including structural, material, and geometric definitions of data items in the design standard. The design knowledge sources are used by various design process modules.

The task specification user interface assists the user in defining the component type, governing standard, design method, design stage, etc., whereas the postprocessor provides the user with commands for displaying information regarding task description, component properties, the constraint set and requirements that were checked, design requirements, etc. The modules in the knowledge base are invoked by the *system controller* based on the current design state, which is represented by messages on the *message blackboard* and design information on the *design information blackboard.* A set of control rules is used by the system controller to specify modules to be invoked based on information present on the message and design information blackboards (Garrett, 1986).

3.6. A PC-BASED EXPERT SYSTEM FOR AUTOMATED REINFORCED CONCRETE DESIGN CHECKING (Saouma et al., 1987)

3.6.1. INTRODUCTION

This expert system (ES) checks reinforced concrete designs based on the ACI 318–83. Several software tools, including the M.1 ES shell (version 2.1 cos), Microsoft Fortran (version 3.30), Microsoft C (version 3.00), and spreadsheet Lotus/123 (Release 2) were used in the development of the system. The overall system architecture is shown in Figure 4.

The system consists of two distribution disks: a "user" disk containing only those files necessary for system operation and a "maintenance" disk containing additional files used in system implementation.

3.6.2. THE SPREADSHEET

The developed spreadsheet (ACI. WK1) is LOTUS/123 compatible and contains the three columns of interest to the user (variable description column A, data entry column B, and legal value column C) and a small data-writing macro in protected cells (column I-O, rows 9 to 14). This macro is used to send spreadsheet data to a datafile for subsequent input to the ES.

3.6.3. SPREADSHEET CONVERSION TO M.1 DATA

An auxiliary file containing internal variable names is used to take output from the spreadsheet and generate an input file for M.1 in a format compatible with the ES input. The sequence for conversion is as follows:

1. Fetch a value from the 123 output file.
2. Fetch the variable name used internally by M.1 for this value.
3. Write the variable and value using the M.1 cache format.

The match between a value in the 123 output file and the internal name used by M.1 is made by using a variable name file. This file contains the internal M.1 variable names in the order in which the values are written from the spreadsheet.

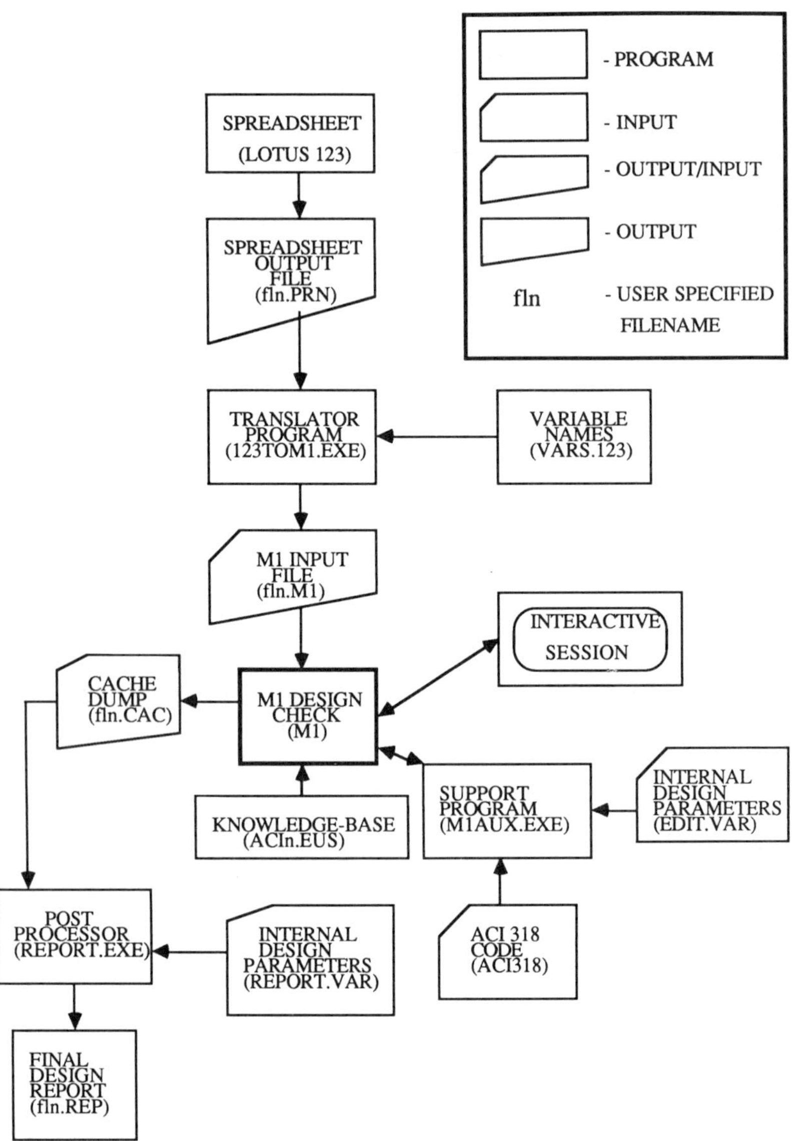

FIGURE 4. System architecture. (From Saouma, V. E., Jones, M. S., and Doshi, S. M., A PC Based Expert System for Automated Reinforced Concrete Design Checking, Rep. Contract DACA39–86-K-0011, U.S. Army Engineer Waterways Experiment Center, Information Technology Laboratory, Vicksburg, MS, 1987. With permission.)

3.6.4. EXPERT SYSTEM

3.6.4.1. The ACI Knowledge Base

The system consists of ten KB files that contain rules pertaining to user interface operation, top-level duties, knowledge base representing Chapter 8 through 12 of the ACI Code (each file has knowledge for an individual chapter, and one file contains knowledge for all the chapters and user interface rules), knowledge base for the beam "quick-check", and knowledge base for the column "quick-check". The design checker will use a different combination of these files, based upon the checking task.

3.6.4.2. Interface Configuration

M.1 provides a menu-driven interface for user interaction, providing a display of system output and menu-driven input. The user can use several ALT keys to issue common commands.

3.6.4.3. External Function

External functions are needed to perform various duties, such as numerical calculation, display of ACI provision text, and cache editing, which cannot be done easily by rules.

3.6.4.3.1. External Code

The source code for the externals is contained in three files. STUFF.C contains the external functions for sine, cosine, cube root, and display of provision text. Data are taken from KB rule via an import statement, calculated in C-library function, and returned to the knowledge base using an export statement. Provision display is obtained by imposing the desired specification number (e.g., 10–3–1) and searching a file containing the ACI code text (CACI 318) for the provision label. Once found, all text up to the first occurrence of "#" (used as a delimiter) is copied to the display. If the specified label is not found, a message stating such is displayed and control is returned to the knowledge base.

This file invokes the column strength external and the editor. The source code for the editor is in EDITOR.FOR, whereas both files are written in FORTRAN. If the current menu screen is saved and cleared by the M.1 function "savescreen", control is passed to the editor subroutine. Here, the current cache is read, and the variables specified in EDIT.VAR are extracted and displayed using the corresponding "user name". The editor now enters an interactive loop, allowing the user to change the rules of the given variables. Upon exit, an updated cache is written by the editor for subsequent input by the KB on external exit. The required input data are imported from the KB, necessary computations are made, and the results stored in a separate file. Control then returns to the C-code, which reads these values and exports them to the KB and returns execution to the ES.

3.6.4.3.2. Rule Partner

Some rules are necessary in the knowledge base to invoke the external code via an "external" statement. In the case of sine, cosine, and cube root, the rule is one of the ACI provisions. The remaining externals are invoked through rules associated with the M.1 command mapped into the appropriate ALT key sequence.

3.6.5. REPORT GENERATOR

The function of the report generator is to extract essential information from a construction cache dump and arrange this information in an aesthestic manner in a report file.

3.7. CONCLUDING REMARKS AND SUGGESTIONS FOR FURTHER WORK

ES applications to generic design standards processing provide an opportunity to represent and make use of requirements and standards in a concise and unambiguous manner, and provide allowable value ranges for undetermined data. The use of codes forms a mandatory requirement in almost all areas of structural design, and hence is a particularly important application area. The emphasis of earlier work in design standards was to improve the representation and organization of standards, analysis, and use of standards. The synthesis of standards is a promising new area for further work.

Chapter 4

CONSTRUCTION ENGINEERING AND MANAGEMENT

4.1. INTRODUCTION

Construction engineering and management can be divided into three major areas: (1) engineering of temporary facilities for construction, (2) management of the construction process, and (3) rehabilitation, repair, and maintenance of facilities. The construction engineering and management involve all the planning and design decisions related to the equipment and physical facilities (e.g., cofferdams, access roads, etc.) involved in the construction process. Expert system (ES) techniques might profitably be applied to (1) the design of construction methods, (2) manufacturing and placing concrete, (3) excavations for construction, (4) constructibility evaluation, (5) site layout, and (6) surveying associated with the precise location of permanent facilities.

The construction management consists of administrative, legal, and financial elements of the construction process. Project planning, scheduling, and control are now widely supported by the use of network-based project scheduling techniques for analysis and by database management systems for reporting. Decisions in contract management include the selection of overall contracting strategy and contract clauses, identification of project financing, selection of prospective contractors or designers, evaluation of progress payments, and potential claims and project organization design. The construction company management consists of marketing strategy decisions, personnel management decisions, company organization design, financial planning, construction equipment policy decisions, and safety management. A number of problems in construction engineering, which has an ill-defined and ill-structured environment, are not amenable to satisfactory solution by procedural, algorithmic computer techniques. The complex nature of the problem requires the knowledge and experience of a recognized expert and several ES have been developed to capture this expertise.

The possible range of expert systems in construction includes (Rehak and Fenves, 1984):

- Interpretation of signals and data from exploratory devices and sensors
- Monitoring performance of equipment and processes
- Diagnosis of equipment failures and process deficiencies
- Recommendations for corrective actions in case of malfunctions and shortages
- Planning of construction activities and equipment functions
- Design of construction schedules

Typical representative ES applications in construction are described in the following subsections (M. Arockiasamy et al., 1989).

4.2. EXPERT SYSTEMS IN CONSTRUCTION ENGINEERING

4.2.1. SOIL EXPLORATION CONSULTANT: SOILCON

The condition of the soil below the surface of the ground is one of the biggest uncertainties in construction projects. The correct assessment of subsurface risk at an early stage of the project can contribute significantly to the overall success of the construction effort. SOILCON eliminates to some extent the uncertainty involved in subsurface exploration by evaluating known conditions of the site and recommending appropriate methods to continue exploration, if required. The system is designed to incorporate subsurface considerations into contract design, thereby reducing contractor contingencies. The output of SOILCON includes a list of recommended investigation procedures ranked by certainty, display of their descriptions, and cost estimates for the methods. The system uses backward chaining from the knowledge base of rules encoded in a PROLOG-like syntax and runs on IBM PC-class computers. It is a developmental ES that does not have the capability to handle quantitative information (Ashley and Wharry, 1985).

4.2.2. LAYOUT OF TEMPORARY CONSTRUCTION FACILITIES: SIGHTPLAN (Tommelein et al., 1987)

4.2.2.1. Introduction

Selection of construction methods and equipment, and the design of the site layout are given attention at the bidding stage and at the startup of construction of the project, but continuous advance planning is seldom carried out. Inappropriate site layout can lead to considerable lost time in the form of excessive travel time of workers and equipment, and inefficiencies. The quality of temporary facilities layout on site has a significant effect on the efficiency, safety, productivity, and cost of construction. The ES SIGHTPLAN is designed to assist project managers in their complex task of designing site layouts and updating the plan continually as project time progresses.

During construction of a project, a number of different temporary facilities are located and removed from the site. Determination of their location is a spatial arrangement problem dealing with positioning objects under constraints. A blackboard ES shell, BB1, has been chosen to apply varying problem-solving strategies and construct the layout incrementally (Figure 1). It is particularly well suited for reasoning about alternative objects, simultaneously searching for multiple hypothetical solutions, and dealing with time. ACCORD is a specialization language that provides a vocabulary to express relationships in spatial arrangements. Objects are assigned roles based on the constraints and with the site.

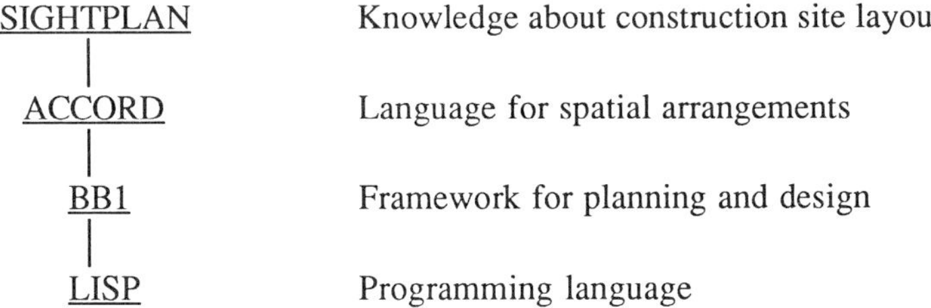

FIGURE 1. SIGHTPLAN software environment. (From Tommelein, I. D., Levitt, R. E., and Hayes-Roth, B., *Managing Construction Worldwide*, Vol. 1, *Systems for Managing Construction*, Lansley, P. R. and Harlow, P. A., Eds., E. & F. N. Spon, New York, 1987, 566. With permission.)

WHELL LOADER	
Type:	930
Length:	5.9 m
Bucket rated capacity:	1 m3
Geometry:	fixed
Mobility:	mobile
Zoning:	any
Duration from:	10Jan88
Duration to:	25Jun88

FIGURE 2. Object description. (From Tommelein, I. D., Levitt, R. E., and Hayes-Roth, B., *Managing Construction Worldwide*, Vol. 1, *Systems for Managing Construction*, Lansley, P. R. and Harlow, P. A., Eds., E. & F. N. Spon, New York, 1987, 566. With permission.)

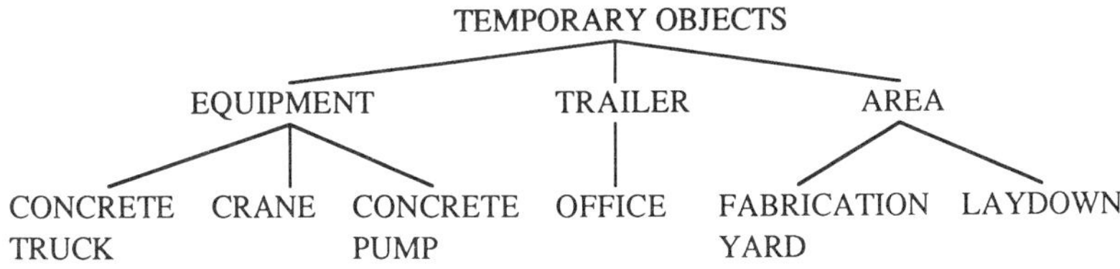

FIGURE 3. Objects in the site layout domain. (From Tommelein, I. D., Levitt, R. E., and Hayes-Roth, B., *Managing Construction Worldwide*, Vol. 1, *Systems for Managing Construction*, Lansley, P. R. and Harlow, P. A., Eds., E. & F. N. Spon, New York, 1987, 566. With permission.)

SIGHTPLAN contains all construction management domain knowledge necessary to design site layouts. Objects are described by their type, dimensions, geometry, mobility, possible zoning requirement, and duration on site (Figure 2). Objects inherit properties from the class to which they belong (Figure 3). The planning mechanism of BB1 allows SIGHTPLAN to propose two or three alternative possibilities to the user that satisfy all or most of the given constraints. The evaluation of a particular design can then be rated by means of a checklist of shortcomings.

4.2.2.2. Example of SIGHTPLAN's Design Actions

A reinforced concrete wall has to be constructed in an excavated area 7 m deep. A crane in the pit lifts reinforcement bars and formwork into place. The current activity is to perform concrete placement by means of crane and bucket. Two subcontractors (subs) are to be involved with the concrete placement: one places reinforcement bars (rebar), the second one sets the formwork. Both subcontractors want to be in the secondary zone (i.e., the zone surrounding the excavation), and within crane reach. Two strategies could apply: (1) place rebar first, then place the formwork around it, or (2) alternately placing one and then the other. Two state families are generated: one for the rebar sub location and one for the formwork sub.

SIGHTPLAN would reason in the following way:

Strategy 1

GOAL — locate sub1 on site
DATA — the area required by sub1 is 900 ft^2 (100 m^2)
CONSTRAINT — the area for sub1 needs to be within crane reach
  ACTION — locate the crane on site; find area of crane reach
    ACTION — find possible areas for sub1—the system finds five possible locations

This set of five possible locations is called a "State Family"
      ACTION — locate sub1 on site—the system decides on one location

Figures 4 and 5 illustrate the five possible locations of sub1 on site. When SIGHTPLAN designs the site at a later stage of construction, the same process will be repeated to locate sub2. Sub1 may decide to take position 1, and when he finishes his work, position 1 is available for use by sub2.

Strategy 2

GOAL — locate both sub1 and sub2 on site
SUBGOAL — locate sub1 on site
DATA — the area required by sub1 is 900 ft^2 (100 m^2)
CONSTRAINT — the area for sub1 needs to be within crane reach
  ACTION — locate the crane on site; find area of crane reach
    ACTION — find possible areas for sub1—the system finds five locations
SUBGOAL — locate sub2 on site
DATA — the area required by sub2 is 900 ft^2 (100 m^2)
CONSTRAINT — the area for sub2 needs to be within crane reach
  ACTION — locate the crane on site; find area of crane reach

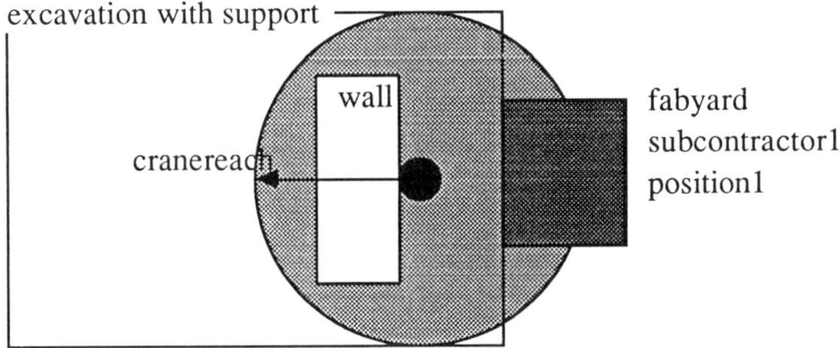

FIGURE 4. Example of reasoning about site layout. (From Tommelein, I. D., Levitt, R. E., and Hayes-Roth, B., *Managing Construction Worldwide,* Vol. 1, *Systems for Managing Construction,* Lansley, P. R. and Harlow, P. A., Eds., E. & F. N. Spon, New York, 1987, 566. With permission.)

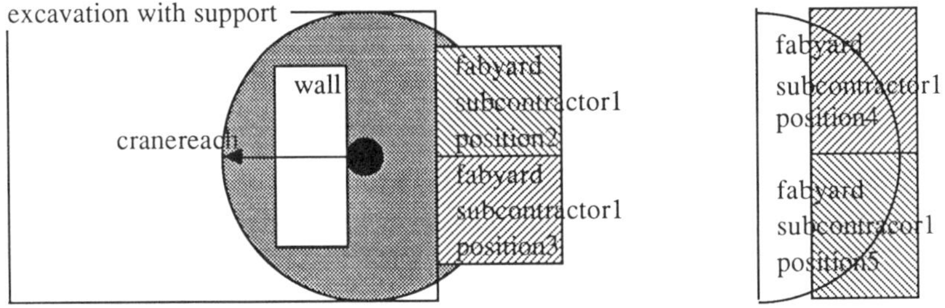

FIGURE 5. Both subcontractors are located on site. (From Tommelein, I. D., Levitt, R. E., and Hayes-Roth, B., *Managing Construction Worldwide,* Vol. 1, *Systems for Managing Construction,* Lansley, P. R. and Harlow, P. A., Eds., E. & F. N. Spon, New York, 1987, 566. With permission.)

ACTION	find possible areas for sub1—the system finds five locations
ACTION	locate sub1 and sub2 simultaneously on site

If the five same possible locations are generated for each sub, several combinations (e.g., sub1 in position 2 and sub2 in position 3) are possible that satisfy their location constraints.

The SIGHTPLAN prototype is built based on a fictitious project and contains intuitive knowledge. It lays out rectangular objects with given geometry and dimensions on an orthogonal site grid. At a given instant of the construction schedule, a limited number of objects is on site. It is proposed to build an *extended system,* and prototype implementation would be refined and elaborated using an existing construction project, field data, and site expertise.

4.2.3. BRICKWORK EXPERT: BERT (Bowen et al., 1986)

BERT is an interactive design aid that evaluates proposed designs for the brickwork cladding of a building. It critically reviews a submitted design from an AUTOCAD system and suggests improvements to the user for editing the drawing.

4.2.3.1. Methodology

The user supplies the design of the brickwork cladding as an input through an IBM PC CAD program called AUTOCAD. This input is then restructured by AUTOCAD's attribute file generator into a text file that describes symbolically the face of the building in question. A graphical representation processor examines the text file and then computes the spatial relationships between the features of the building. Rules about the proper location of the movement joints are incorporated into the knowledge base of the system, and then mapped into LUCIFER programming language rules. The main architecture of LUCIFER is based on forward chaining, although there are provisions for backward chaining and a blackboard-type architecture, enabling the knowledge from LUCIFER to be shared by other ES. BERT also has a brick database that stores details about the parameters of each of the different types of bricks. After completing an analysis of the design, BERT will recommend changes in the design that may be incorporated by the user into the original design. The user may then resubmit it to BERT for another cycle, or exit the program. BERT is an operational prototype ES designed in conjunction with a major brick manufacturer in order to standardize design advice to architects.

4.2.4. SITE LAYOUT EXPERT SYSTEM: CONSITE (Hamiani and Popesen, 1988)

4.2.4.1. Introduction

CONSITE has been developed to demonstrate the viability of the knowledge-based expert system (KBES) approach to the jobsite layout problem. Its knowledge base contains representations of the site and the temporary facilities to be located, and also embodies the design knowledge of the expert. It manipulates the facilities, extracts information from the actual site layout, generates alternative locations for the facilities, tests constraints, selects a location, and updates the layout. CONSITE uses a representation that is a mixture of rules, frames, and object-oriented programming in the KEE environment.

4.2.4.2. Facility and Site Representation

The site is divided into a set of convex polygons of three possible types: open, closed, or access. The open space is the space available for facilities location, whereas the closed space is that already used up by any kind of obstruction such as trees, existing buildings, etc. The access space is the space needed by workers and equipment at the site. Each of the polygons is further made up of a set of sides, each side being unique and part of only one polygon. Figure 6 shows the representative job site of an office building in CONSITE. A convex polygon representation of a job office is illustrated in Figure 7.

4.2.4.3. Design Knowledge and Design Status Representations

The expert's design knowledge consisting of heuristic and rules of thumb acquired through years of experience is represented in CONSITE as a set of rules. These rules recognize the commonly occurring patterns of layout by identifying the facility, extracting information from the actual layout, activating methods that generate possible locations for the facility at hand, and updating the layout representation.

Design status and related information are monitored by CONSITE using a frame named Design that has attributes whose values change in time to represent the different states of the layout process. This unit keeps track of the facility being located, the alternative locations generated, and the alternative selected at the previous level of the design process. It also keeps a list of the polygons that represent the site at the current stage of the design.

4.2.4.4. Alternative Representation

During the design process, CONSITE generates alternative locations for the facility to be entered into the layout. These alternatives are generated as frames with attributes that allow their identification and evaluation. The set of constraints represented in CONSITE forms an important set of attributes.

4.2.4.5. Constraints Representation

Constraints in CONSITE are desired qualities of the layout due to relationships between the facilities and the work area, the facilities and the outside world, or between the facilities themselves. Interaction of facilities with other facilities, the work area, or the region outside the site boundaries affects their location. The constraints implemented in CONSITE are (1) adjacency, (2) distance, (3) access, (4) spatial, (5) position, and (6) view.

4.2.4.6. Knowledge Base Organization

The knowledge base organization is shown in Figure 8, and the knowledge is represented in frames and rules. The frames define the static knowledge that represents objects in the layout and their attributes, which are either descriptive or procedural in form, and allow a description of real objects such as the site and the facilities, and of abstract objects such as the polygons, sides, and points. Rules represent both heuristic and judgmental reasoning knowledge, whereas the object-oriented programming describes procedural language, such as numerical processing, overlay checking, translation, and rotation of the facilities. This data-dependent programming is attached to slots in frames describing specific objects.

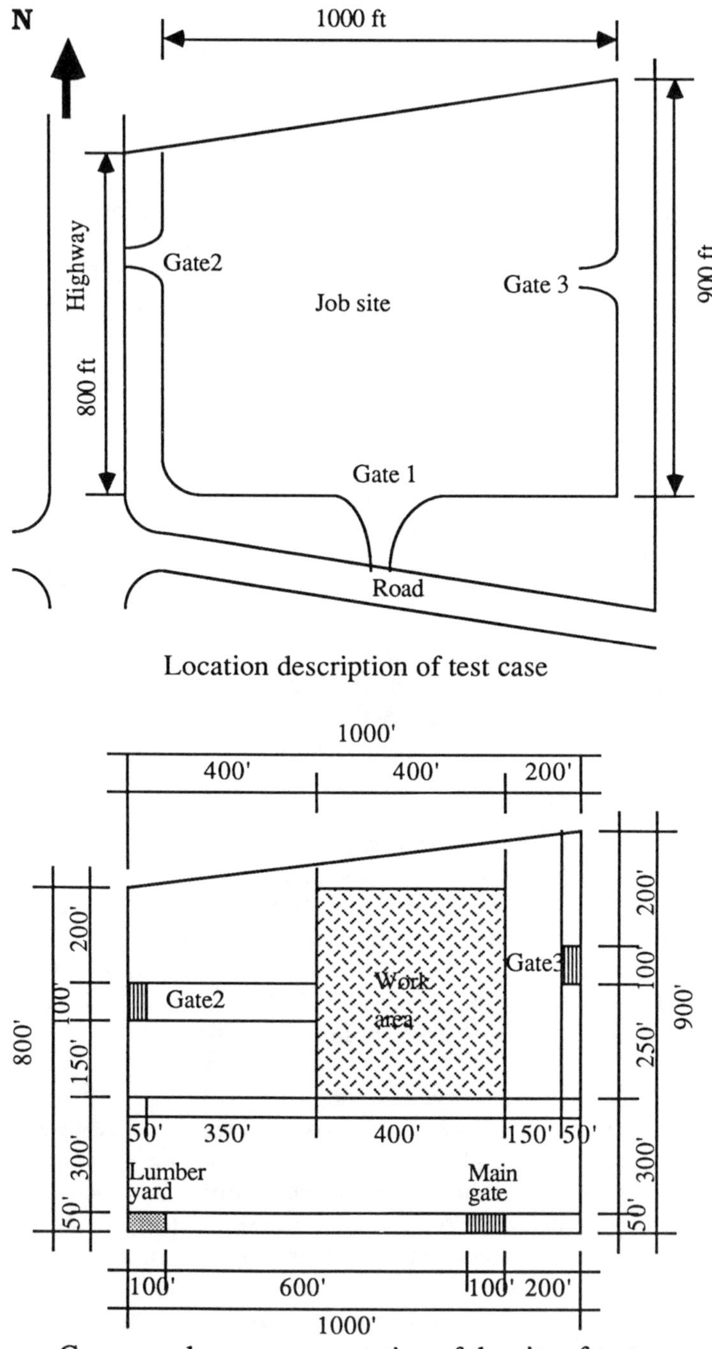

FIGURE 6. Convex polygon representation of a job office. (From Hamiani, A. and Popesen, C., Proc. 5th ASCE Conf. Computing in Civil Engineering: Microcomputers, Will, K. M., Ed., ASCE, New York, 1988, 248. With permission.)

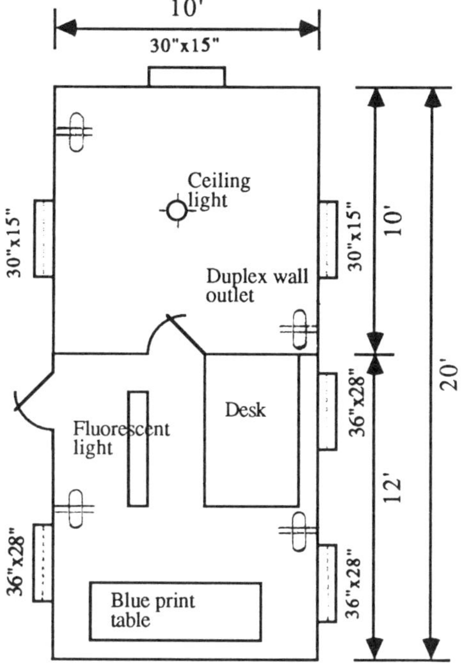

Engineering drawing of job office

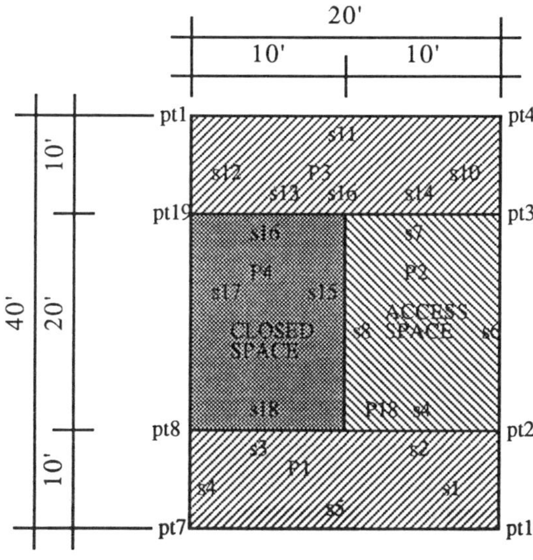

Spatial requirement of job office

FIGURE 7. Representation of the job site of an office building. (From Hamiani, A. and Popesen, C., Proc. 5th ASCE Conf. Computing in Civil Engineering: Microcomputers, Will, K. M., Ed., ASCE, New York, 1988, 248. With permission.)

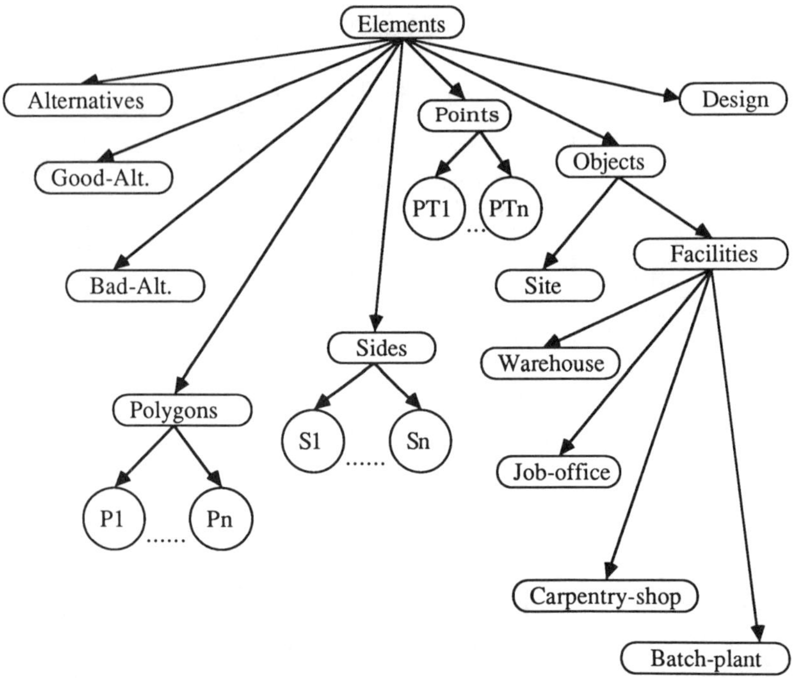

FIGURE 8. Organization of the knowledge base. (From Hamiani, A. and Popesen, C., Proc. 5th ASCE Conf. Computing in Civil Engineering: Microcomputers, Will, K. M., Ed., ASCE, New York, 1988, 248. With permission.)

4.2.4.7. Problem-Solving Strategy

CONSITE uses a plant-generate-test strategy. The planning phase takes advantage of the order in which the expert enters the facilities into the layout. This ordering is implemented through task sequencing. At every level of the search tree, the task is to find a location for the actual context (facility). Once the context is known, CONSITE activates the generator, which is a set of LISP functions that manipulate the space representation and generate alternate locations. The selection of the alternative is done during the testing phase after checking the facility location for the constraints and transferring them to the good-alternative or bad-alternative category. The final location is selected from the good alternatives and implemented through an update of the list of polygons representing the layout in the frame design. The output is displayed graphically on the screen, indicating the location of each facility on the site. The output from CONSITE after solving the office building problem is shown in Figure 9.

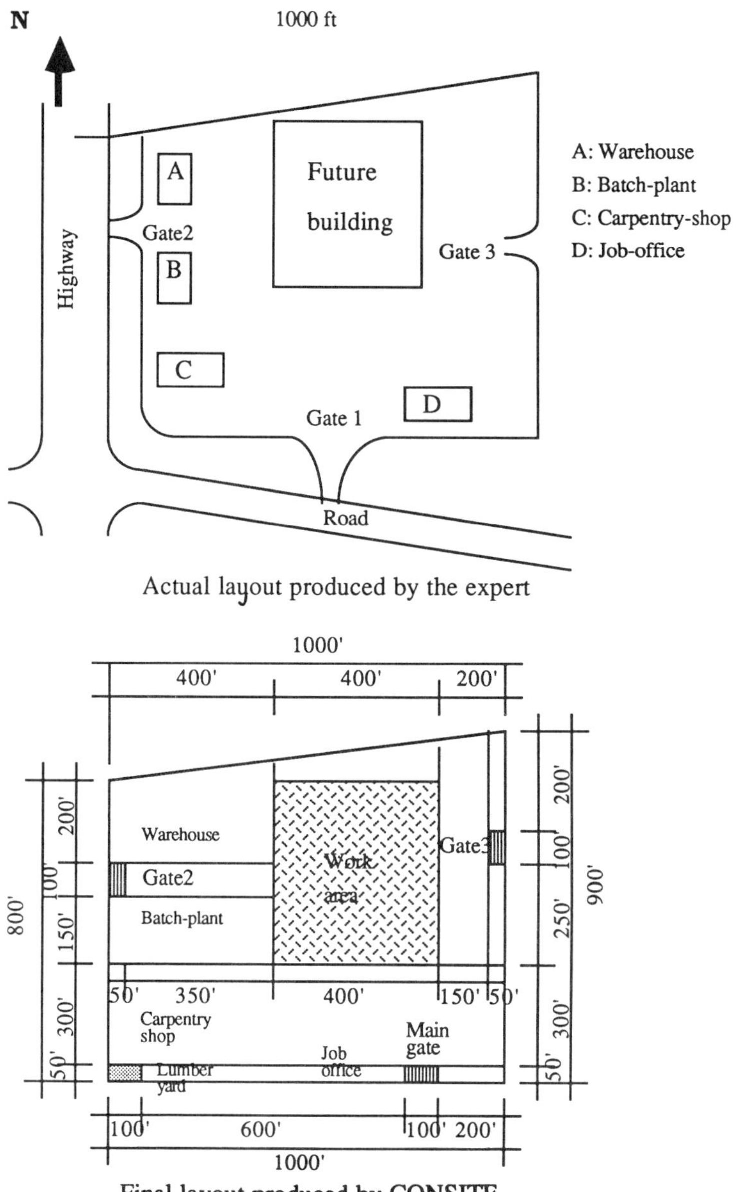

FIGURE 9. Output of CONSITE after solving the office-building problem. (From Hamiani, A. and Popesen, C., Proc. 5th ASCE Conf. Computing in Civil Engineering: Microcomputers, Will, K. M., Ed., ASCE, New York, 1988, 248. With permission.)

4.2.5. EXPERT SYSTEM FOR CONTRACTOR PREQUALIFICATION (Russell and Skibniewski, 1988)

4.2.5.1. Introduction

A prototype rule-based expert system is being developed to aid in the contractor prequalification decision-making process from an owner's perspective. The task of selecting the "right" bidder for a particular project is one of the most challenging tasks performed by an owner or contract administrator. Contractor prequalification is a decision-making process involving a wide range of criteria for which information is often qualitative and subjective.

4.2.5.2. Knowledge Acquisition Strategy

The knowledge acquisition process involved the following three steps: (1) gathering general information (viz., identification of decision factors and subfactors, peculiarities, and biases in the process) on the prequalification process, (2) development of a questionnaire on the impact of major decision factors and subfactors on the prequalification decision-making process, and (3) structuring the subfactors into sub-subfactors, and extracting, formalizing, and developing qualitative and quantitative rules.

4.2.5.3. Knowledge Base Design

The structure of the knowledge base presented in Figure 10 consists of two modules:

Decision-maker module (owner)—This module represents the characteristics of the decision maker (owner) that impact the selection of the decision strategy and the development of the prequalification criteria.

Contractor module—This module is used to store appropriate characteristics of the contractors being prequalified.

The characteristics of the decision maker include, among others, items such as type of owner (e.g., public or private), owner objectives, type of construction, and contracting strategy. The decision strategy selected can include dimensional weighting, a two-step prequalification process, and subjective judgement. Table 1 presents the major composite factors relevant to the decision-making process for public owners and private owners/construction managers. Each of the composite factors listed in Table 1 can be further characterized by the factors that make up the given factor.

For example, the "management" composite factor for private owners/construction managers consists of:

- Project control procedures
- Project management capabilities
- Staff available
- Company organization

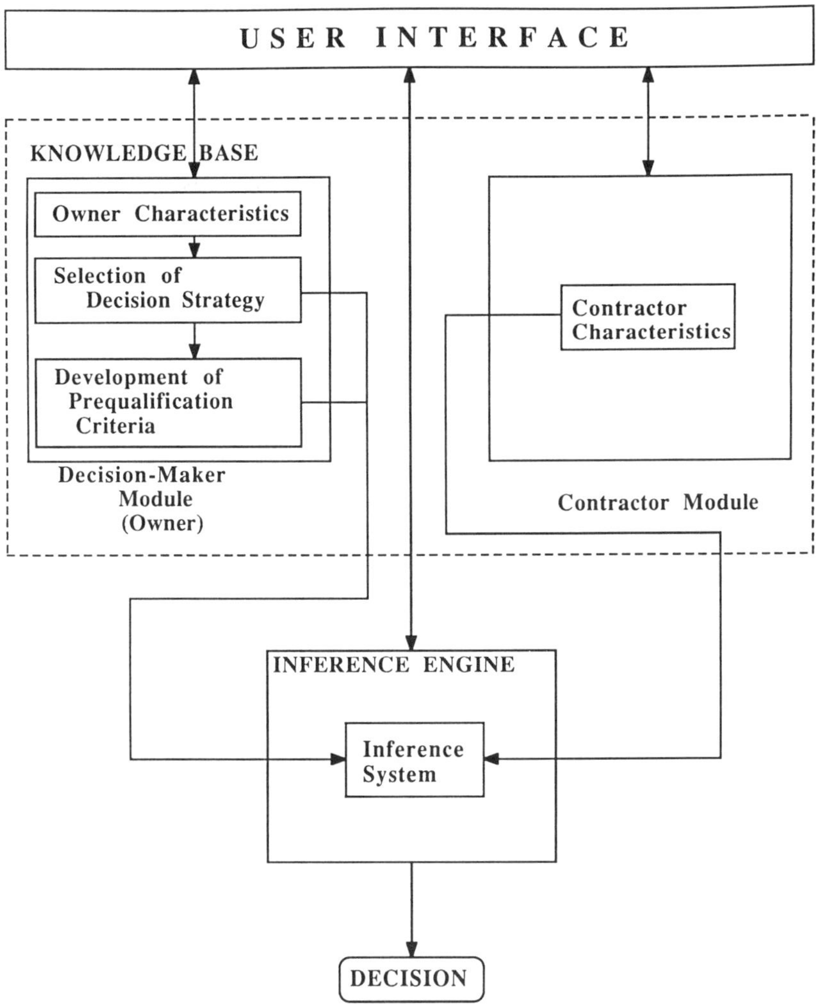

FIGURE 10. Structure of a rule-based expert system for contractor prequalification. (From Russell, J. S. and Skibniewski, M. J., Proc. 5th ASCE Conf. Computing in Civil Engineering: Microcomputers, Will, K. M., Ed., ASCE, New York, 1988, 248. With permission.)

The factors can be further characterized by subfactors. For example, company organization consists of:

- Type of ownership (e.g., partnership, corporation, sole owner)
- Number of years in construction
- Contractor's licenses held (by state and/or by type of work)
- Number of times a contractor has failed to complete a contract
- Appropriateness of company organizational structure

TABLE 1
Major Factors Utilized in the Contractor Prequalification Process

Group	
Public owners	**Private owners and construction managers**
Performance	Management
Type of contractor	Safety
Capacity for New York	Location
Location	Performance
Worked performed percentage	Resources
Third-party evaluation	Financial stability and experience
Financial stability	Failed performance
	Bonding capacity
	Capacity for New York

The financial stability factor can be broken down into four subfactors:

- Credit rating
- Banking arrangements
- Bonding
- Balance sheet

The balance sheet subfactor can be further reduced to the following parameters:

- Net worth (shareholder's equity)
- Working capital
- Debt/net worth ratio

The knowledge will be represented by self-contained pieces of knowledge in the form of "if . . . then" production rules. The standard syntax adopted for a production rule is

IF (condition)
THEN (action)

At each level of the hierarchy, production rules must be formulated. Table 2 shows a sample production for the balance sheet subfactor.

The contractor prequalification expert system (ES) will be developed utilizing a backward-chaining inference mechanism. The procedure will be invoked to determine whether each contractor is acceptable to submit a bid for the project.

TABLE 2
Sample Rules for the Subfactor "Balance Sheet"

IF	Working capital <\$0.00 (current assets − current liabilities)
THEN	Contractor is experiencing cash flow difficulties and the contractor's banking arrangements should be checked
IF	Debit/net worth ratio >3:1 (shareholders' equity)
THEN	Contractor is highly leveraged and is not carrying a majority of the financial risk (the bank and/or material suppliers and/or equipment suppliers are carrying the risk)
IF	Net worth <\$10,000 (shareholders' equity)
THEN	Contractor does not have enough financial risk. In the event of an unforeseen situation (e.g., loss of money on the project), it is highly likely the contractor will not stay and complete the project
IF	The amount (\$) in common capital stocks <30% of net worth (shareholders' equity)
THEN	Shareholders have little equity in the business
IF	Working capital <\$0.00 and net worth <\$100,000 and banking arrangements = no
THEN	Contractor cannot pay his bills
IF	Debt/net worth ratio >3:1 and net worth <\$10,000
THEN	Contractor currently does not have substantial financial risk to guarantee the completion of the project

4.2.6. EXPERT SYSTEMS IN REAL-TIME CONSTRUCTION OPERATIONS (Paulson and Sotoodeh-Khoo, 1987)

4.2.6.1. Introduction

The optimum loading time for an earthmoving scraper varies with the length of the haul, as well as with changes in variables, such as the current soil grain size, moisture content, cohesion, and density. Less-experienced operators may overload or underload their scrapers under rapidly varying conditions. The ES are designed to specify a fleet of equipment for a given project, aid new operators to understand optimum loading times for each machine, and optimize fleet production by communicating between machines in real time.

Real-time data collection via electronic instrumentation of construction field operations can be joined with knowledge-based expert systems (KBES) to implement analytical modeling procedures such as simulation and nonlinear production optimization. The real-time instrumentation and monitoring of earthmoving scraper operations have been interfaced with the EXSYS ES shell running on an IBM PC/AT computer for implementing the nonlinear optimization method.

4.2.6.2. Methodology

4.2.6.2.1. Single-Scraper Expert System

The scraper operator inputs the type of machine, the type of soil (e.g., sand, clay, etc.), and the working conditions (e.g., wet ground, low traction, etc.) to be the on-board ES. Based on the input information, the knowledge base would inform the operator of a range of load times where he is most likely to achieve maximum production. A load time from the specified range (the middle value of the range, for instance) would be selected by the on-board monitoring system. It warns the operator to stop loading and start hauling as soon as that specified load time has elapsed. A few seconds into the haul cycle, the load cells mounted on the machine could determine the average payload based on several samples of the payload. The balance of cycle time (i.e., haul, dump, and return times) for the first run may be provided from the knowledge base, and those in subsequent runs computed accurately by recording data from load sensors, strain gages, gravity mass sensors, optimal volume sensing, inertia sensing, gearbox sensing, and speedometer readings. The machine production in volume units per hour for a specific load time and a known cycle time and the payload could be computed and stored in memory for later reference and comparison.

The system would pick a different load time (either higher or lower than the first one) on subsequent loadings, and production per hour could again be calculated based on the payload during the haul cycle for this load time and compared to the previous one. An increase in production would mean that the load time is approaching the optimum value. On the other hand, with a decrease in production, the system would try a different load time in the opposite direction. Within a few iterations, the system would ultimately converge on a load time close to the optimum based on the inference of the appropriate rules; the operator would be advised to operate at that load time until a different value was obtained by the system due to a change in one of the factors affecting production (e.g., an increase or decrease in the cycle time or a change in material properties affecting load time, in the equipment fleet, in haul road conditions, etc.). Figure 11 shows a succession of such points that define a portion of the actual load-growth curve.

4.2.6.2.2. Fleet Management Expert System

The coordination of a fleet of earthmoving machines consisting of the same size and type or of differing sizes becomes more complex and challenging to minimize wait times for pushers and scrapers during their respective cycles. The fleet optimization problem is carried out using rule-based logic in the EXSYS environment. This software not only allows deduction using rule-based logic in achieving a theoretical optimum fleet balance and load time for each scraper, but also checks and adjusts this theoretical value based on real field data collected through the on-board sensors.

The knowledge base would compute the correct number of machines to achieve the completion goal based on user input information about the project duration and the earthmoving volume. Knowing the haul and return road lengths, grades, and rolling resistance, the knowledge base would access external data bases to calculate the scraper travel speeds (when loaded and empty) and determine a theoretical time for the balance of the cycle. Using additional logic based on the fleet theory, a load time is then selected such that scraper and pusher times are optimally balanced. The selected theoretical

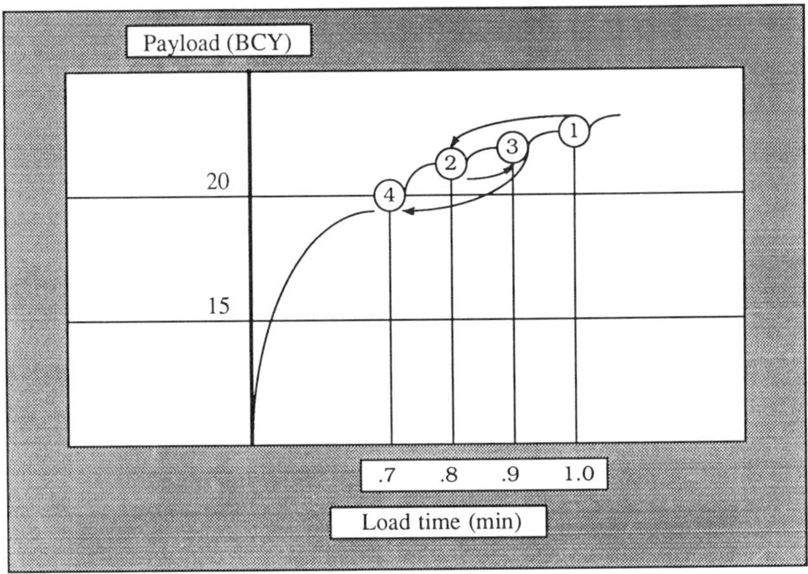

FIGURE 11. Succession of points defining a portion of the load-growth curve. (From Paulson, B. C., Jr. and Sotoodeh-Khoo, H., *Managing Construction Worldwide*, Vol. 1, *Systems for Managing Construction*, Lansley, P. R. and Harlow, P. A., Eds., E. & F. N. Spon, New York, 1987, 566. With permission.)

load time is then transmitted to the field for validation in field conditions. The communication between the ES and the real-time data acquisition program for data validation would eventually enable the logic-based program to learn from its past suggestions and make better decisions in the future under similar job and equipment conditions.

4.3. EXPERT SYSTEMS IN CONSTRUCTION MANAGEMENT

4.3.1. INTRODUCTION

Construction management includes planning, scheduling, and control of construction activities as well as the design of legal, behavioral, and other elements of the construction process. Potential applications of ES in the area of construction project monitoring involves checking, regulating, and controlling the performance and execution of the project. Only selective ES applications in construction management are presented in the following section.

4.3.2. EXPERT SYSTEM ARCHITECTURE FOR CONSTRUCTION PLANNING: CONSTRUCTION PLANEX (Fenves et al., 1988)

Construction planning involves the choice of construction technologies, definition of work tasks, estimation of required resources, durations and costs, and preparation of project schedules. CONSTRUCTION PLANEX is a KBES which synthesizes activity networks, diagnoses resource needs, and predicts durations and costs. The system will generate a plan automatically, or a planner can review and modify decisions during the planning process. The system has three essential parts, as illustrated in Figure 12. The *context* stores information on the particular project being considered, including the design, site characteristics, planning decisions made, and the current project plan. The *operator module* contains operators which create, delete, or modify the information stored in the context. Operators are of two types: *specialized* and *control*. Specialized operators are used for tasks such as technology choice, activity synthesis, duration estimation, etc. The order in which specialized operators are executed is determined by control operators. Interaction between the two types of operators occurs by means of a message interface representing the role of a blackboard. The *knowledge base* contains distinct *knowledge* sources of tables and rules specific to particular technology choices, activity durations, or other considerations. Each knowledge source is used by a particular operator. A user interface with an explanation module is included in addition to the central components.

The following variety of objects storing information in the context are available (Hendrickson et al., 1987):

- *Design element* objects that store information about design components
- *Quantity take-off* objects that store information about elements of work
- *Site-characteristic* objects that store information about different conditions on site
- *Activity* objects that represent construction tasks at different levels of aggregation
- *Resource* objects indicating the characteristics of equipment, labor, or materials
- *Goal* objects that define different stages in the planning process
- *State* objects used dynamically to describe the characteristics of the planning process
- *Constraint* objects to represent required relationships among states and variables
- *Decision* objects for representing points in the planning process that are affected by technology choice, resource allocation, or other decisions made by the user or CONSTRUCTION PLANEX
- *Explanation* objects to store information or pointers to information about the construction plan

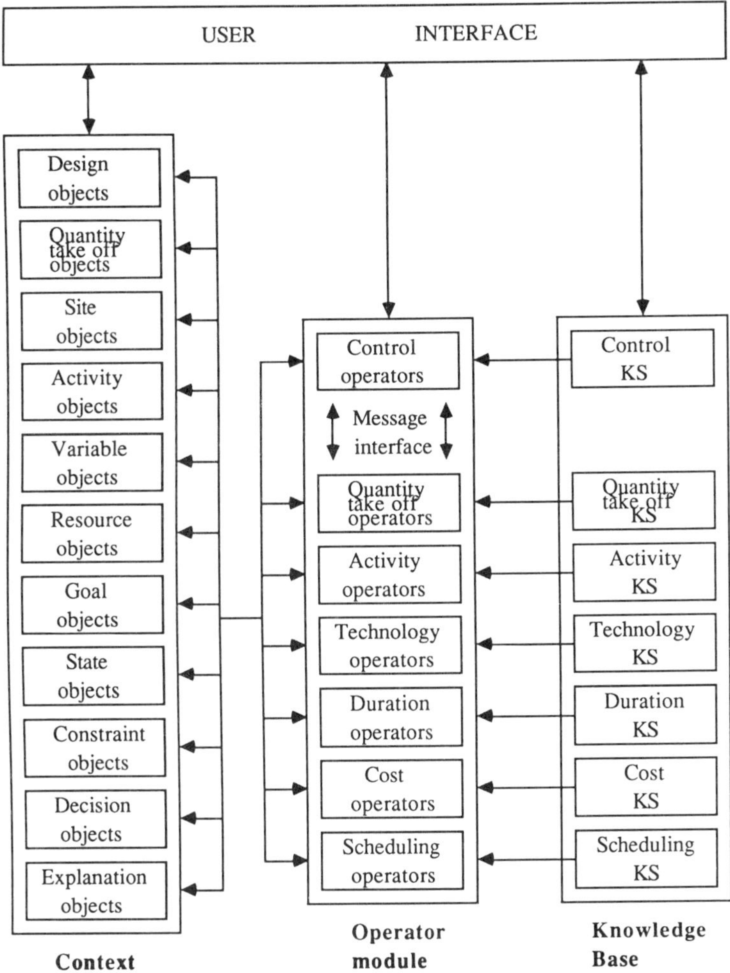

FIGURE 12. Overview of CONSTRUCTION PLANEX. (From Fenves, S. J. et al., *Computer Aided Design*, Vol. 22, No. 1, Butterworth, Guildford, England, 1990, 27. With permission.)

The above mentioned objects are related by a network of relations representing the current project plan, decisions made during the planning process, and different joining schemes. The set of activities thus forms a project network, whereas the system context contains a more extensive network that also registers the planning process and other information. The generation of elements of work defined by the user in the prototype is automated by the insertion of design element objects in the context. Typical modules contained in the operator module are the following:

- *QTO* operators to create elements of work based on design element information
- *Activity* operators to create, elaborate, expand, link or aggregate activities
- *Technology* operators to suggest appropriate equipment or technology
- *Duration* operators to perform estimation
- *Scheduling* operators to provide a project schedule including critical path identification and any required resource allocation

All operators are generic, so that a single operator can be used for all activities. For example, the duration estimation operator would be called for each element activity and a knowledge source specific to that activity consulted to obtain a duration estimate.

The operation of CONSTRUCTION PLANEX relies heavily on a number of distinct *knowledge sources* (Zozaya, 1987). An example knowledge source applied to a small task in the overall planning process is illustrated as a decision table in Figure 13. Different sets of activities required for the construction of a footing are suggested in the knowledge source depending upon soil characteristics. CONSTRUCTION PLANEX knowledge sources perform as small expert systems themselves by supporting numerical functions, calls to other knowledge sources and binding.

PLANEX performs the following sequence of operations in the initial creation of a construction plan:

- *Create element activities* for design elements. A set of element activities required to construct each design element about precedences among activities, technologies to employ, required resources, etc. are made by other operators.
- *Group element activities* of common characteristics in order to have a hierarchy of element activities similar to that of MASTERFORMAT. Thus, element activities are associated with particular physical design elements (such as a column or a beam) and aggregations of activities called *project activities* and *project activity groups.*
- *Determine amounts of work* for element activities. Geometric information for the quantity take-off is inherited from design element frames in the central data store.
- *Select units of measure* for element activities. Crew productivities or material quantities may be expressed in different units (e.g., days instead of hours).
- *Determine material packages* for element activities based on design specifications.
- *Create project activities* that aggregate element activities and provide summary information on the underlying element activities.

Name : KS-Example			Type : first			
Object	Slot	Op	Value	RULES		
current-object	type-element	is	cast-in-place concrete column-footing	t	t	f
soil-characteristics	backfill	is	yes	t	f	i
excavate-column-footing				x	x	i
dispose-excavation-column-footing				x	x	i
pile-up-excavation-column-footing				x	i	i
boorow-material-column-footing				i	x	i
place-forms-column-footing				x	x	i
reinforce-column-footing				x	x	i
pour-concrete-column-footing				x	x	i
remove-forms-column-footing				x	x	i
KS-other-elements				i	i	x

FIGURE 13. Illustration of a CONSTRUCTION PLANEX knowledge source. (From Fenves, S. J. et al., *Computer Aided Design,* Vol. 22, No. 1, Butterworth, Guildford, England, 1990, 27. With permission.)

- *Determine precedences* for project activities. Scheduling is performed at the project activity level, reflecting the homogeneity of resource use and the small granularity of detail contained in the underlying element activities in CONSTRUCTION PLANEX.
- *Compute lags* for project activities. Element activities of several project activities are structured into an element activity subnetwork. Relevant lags among project activities based on this subnetwork is determined using a critical path algorithm.
- *Select technologies* for project activities. Technologies are chosen at a macroscopic or project level, since consistency in this regard will reduce costs.

- *Estimate durations* for project and element activities.
- *Schedule* project activities using CPM, resource allocation, and constraint satisfaction.
- *Estimate costs* by computing activity costs and project costs using unit costs and scheduling information.

The CONSTRUCTION PLANEX system could be applied to different types of projects, although each type would require different knowledge sources. The system is now being implemented in the KNOWLEDGE CRAFT expert system (ES) environment for the application domain of office building construction.

4.3.3. ANALYSIS OF CONTINGENCIES IN PROJECT PLANS: PLATFORM-II (Kunz et al., 1986)

PLATFORM-II is an ES developed to illustrate the use of the artificial intelligence (AI) technique of "multiple worlds" in making project feasibility decisions under uncertainty. This technique assists the project manager in making a decision involving multiple uncertainties by generating "worlds" that describe all the possible combinations of choices available to the project manager together with the implications of those choices and their outcome probabilities and values based on user-specified evaluation criteria.

4.3.3.1. Methodology

PLATFORM-II was developed using the Intelli Corporation Knowledge Engineering Environment (KEE) and employs the frames, rules, and graphics that are integrated in KEE. The use of the assumption-based truth maintenance system (ATMS) of KEE, Version 3, is a significant feature of PLATFORM-II. The user is allowed to make assumptions regarding a decision (e.g., whether to choose to build the graving dock for construction of the concrete base of a platform in Norway or Scotland). Project cost and duration are dependent upon decisions that must be made by the project manager. Figure 14 illustrates part of the frame structure of the application knowledge base (KB) PLATFORM-II. The PLATFORM-II KB uses units to represent diverse objects such as individual activities in the project schedule, rules to update the schedule, and graphical images on display panels. The GEOLOGICAL.ALTERNATIVES unit heads a subtree that describes the geological conditions likely to be encountered in the building of the graving dock. The LABOR.PRODUCTIVITY.ALTERNATIVES unit forms a subtree describing the labor productivity that must be taken into account in building the platform. The facts for each scenario or world combine knowledge about the probabilities for each geological and labor productivity alternative at that site.

Rule premises identifying the project and its location as issues are shown in Figure 15. Alternatives are specified in the rule premises as the project and location units belonging to the referenced class units, viz., the DRILL-

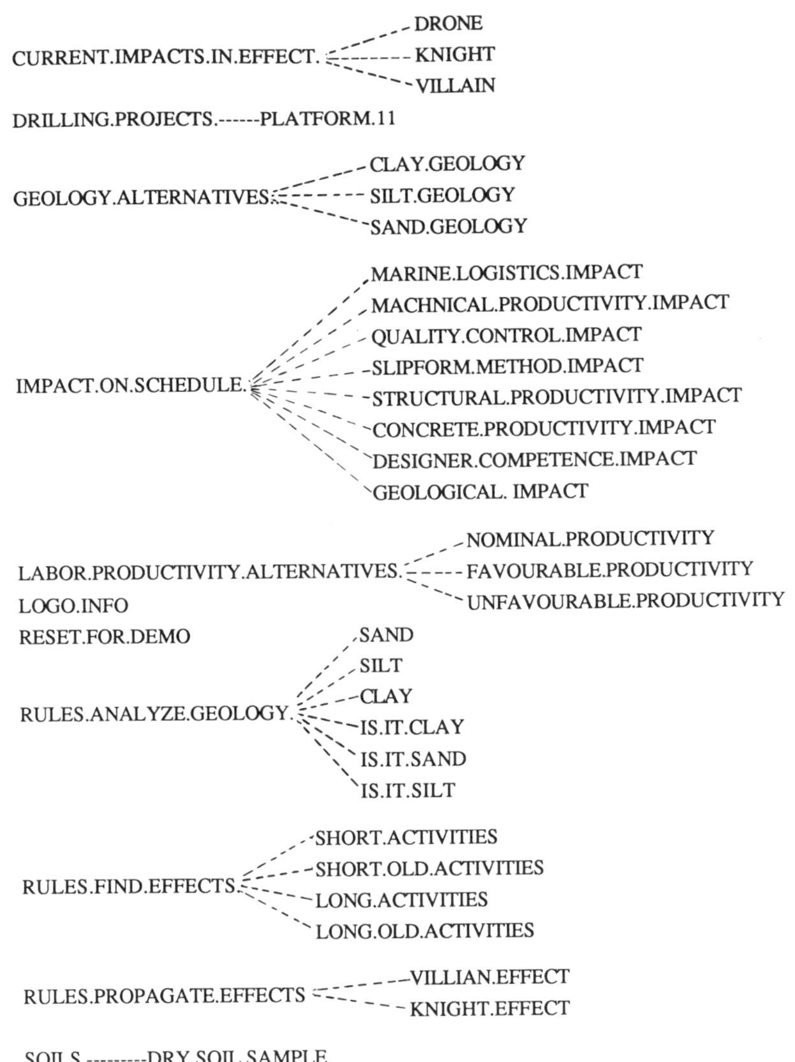

FIGURE 14. A portion of the Platform-II project knowledge base. (From Kunz, J. C., Bonura, T., and Stelzner, M. J., *Proc. First International Conference, Southampton University, U.K.*, Vol. 2, Sriram, D. and Adey, R. A., Eds., Springer-Verlag, New York, 1986. With permission.)

ING.PROJECTS and POSSIBLE.DOCK.LOCATIONS units. The THEN portion of the rule specifies the appropriate conclusion to make for a given set of issues and alternatives. The rule conclusion records the site and a set of likelihoods for each different location in which the project might be constructed.

```
(IF (?PROJECT IS IN CLASS DRILLING.PROJECTS) AND
   (?LOCATION IS IN CLASS POSSIBLE.DOCK.LOCATIONS))
 THEN CREATE.WORLD
     (THE SITE OF ?PROJECT IS ?LOCATION)
     (THE LIKELIHOOD OF NOMINAL.PRODUCTIVITY OF ?PROJECT
          IS (THE LIKELIHOOD OF NOMINAL.PRODUCTIVITY OF ?LOCATION))
     (THE LIKELIHOOD OF FAVORABLE.PRODUCTIVITY OF ?PROJECT IS
        (THE LIKELIHOOD OF FAVORABLE.PRODUCTIVITY OF ?LOCATION))
     (THE LIKELIHOOD OF UNFAVORABLE.PRODUCTIVITY OF ?PROJECT IS
        (THE LIKELIHOOD OF UNFAVORABLE.PRODUCTIVITY OF ?LOCATION))
     (THE LIKELIHOOD OF SAND.GEOLOGY OF ?PROJECT IS
        (THE LIKELIHOOD OF SAND.GEOLOGY OF ?LOCATION))
     (THE LIKELIHOOD OF SILT.GEOLOGY OF ?PROJECT IS
        (THE LIKELIHOOD OF SILT.GEOLOGY OF ?LOCATION))
     (THE LIKELIHOOD OF CLAY.GEOLOGY OF ?PROJECT IS
        (THE LIKELIHOOD OF CLAY.GEOLOGY OF ?LOCATION))
```

FIGURE 15. Rule identifying the project and its location as issues. (From Kunz, J. C., Bonura, T., and Stelzner, M. J., *Proc. First International Conference, Southampton University, U.K.*, Vol. 2, Sriram, D. and Adey, R. A., Eds., Springer-Verlag, New York, 1986. With permission.)

```
(IF  (?PROJECT IS IN CLASS DRILLING.PROJECTS)
     (?SOME.GEOLOGY IS IN CLASS GEOLOGY.ALTERNATIVES)
     (?LABOR.PRODUTIVITY IS IN CLASS
            LABOR.PRODUCTIVITY.ALTERNATIVES)
     (?SELECTED.LOCATION IS IN CLASS DOCK.LOCATION.ALTERNATIVES)
  THEN CREATED.WORLD
     (THE RESULT.OF.GEOLOGICAL.EXPLORATION OF ?PROJECT IS
            ?SOME.GEOLOGY)
     (THE LABOR.PRODUCTIVITY OF ?PROJECT IS
            ?LABOR.PRODUCTIVITY)
     (THE LOCATION OF ?PROJECT IS ?SELECTED.LOCATION)
     (THE COST OF ?PROJECT IS '(COMPUTE.PROJECT.COST $WORLD$))
     (THE DURATION OF ?PROJECT IS
            '(COMPUTE.PROJECT.DURATION $WORLD$)))
```

FIGURE 16. Rule used to identify the issues and alternatives in the problem analysis. (From Kunz, J. C., Bonura, T., and Stelzner, M. J., *Proc. First International Conference, Southampton University, U.K.*, Vol. 2, Sriram, D. and Adey, R. A., Eds., Springer-Verlag, New York, 1986. With permission.)

Figure 16 shows the rule that stipulates the problem to be analyzed. The issues are recognized in the premises-drilling projects, geology, labor productivity, and siting. The CREATE.WORLD operator in the conclusion forms a new world for each situation in which the premises are valid, and it specifies the conclusion to make a new world. In this example of building the dock, there is one drilling project, three geological alternatives, three labor productivity alternatives, and two location alternatives; hence, an exclusive set of 18 different worlds is created by the CREATE.WORLD operator. The conclusion part of the rule asserts values of the named attributes in each world, such as LOCATION and COST. The location is determined from the premises, and the cost computed by a cost function created by the user. Each world is available for inspection by the reasoning rules and by the interactive explanation system. If a line of reasoning becomes inconsistent with earlier assumptions, PLATFORM-II backtracks until it can find an appropriate place to modify the search tree. The user may modify assumptions at any time and let the system generate new worlds. Multiple worlds permit rapid computation of outcome values and allow users to easily create new worlds with slightly different facts and examine their impact on the decision, or to indicate that certain worlds are inconsistent with specified criteria. PLATFORM-II analyzes cost and time outcomes for each of the worlds generated using a complex PERT model with 50 to 100 activities and a realistic cost function that takes into account direct and indirect costs, including time-related bonus/penalty amounts. PLATFORM-II, which is an operational ES, is currently used to demonstrate the ATM capabilities of KEE.

4.3.4. KNOW-HOW TRANSFER METHOD (Niwa and Okuma, 1982)

The know-how transfer method is intended to improve engineering or project management. The dramatic changes in the world economic balance in the 1970s led to many large construction projects in the Middle East. These projects faced long delays in implementation resulting from problems associated with working within a different culture, with different social and religious values. The know-how transfer method was designed to help project managers with risk management at the project execution stage, and its main focus is risk identification.

4.3.4.1. Methodology

The basic feature of this ES is the development of the know-how transfer method of acquiring knowledge for the system to use. Multidisciplinary knowledge in the different areas of managerial, technical, economic, financial, social, scientific, legal, and political skills constitutes the know-how. The system stores the risk know-how onto a standard work package matrix (Figure 17). The standard work package matrix consists of columns indicating activities and rows indicating objects. Each job in the project is an intersection of an activity and an object. Know-how acquired on a project is also related to an activity and an object, and then placed onto the grid. This know-how grid is subsequently mapped onto the standard work package matrix so that

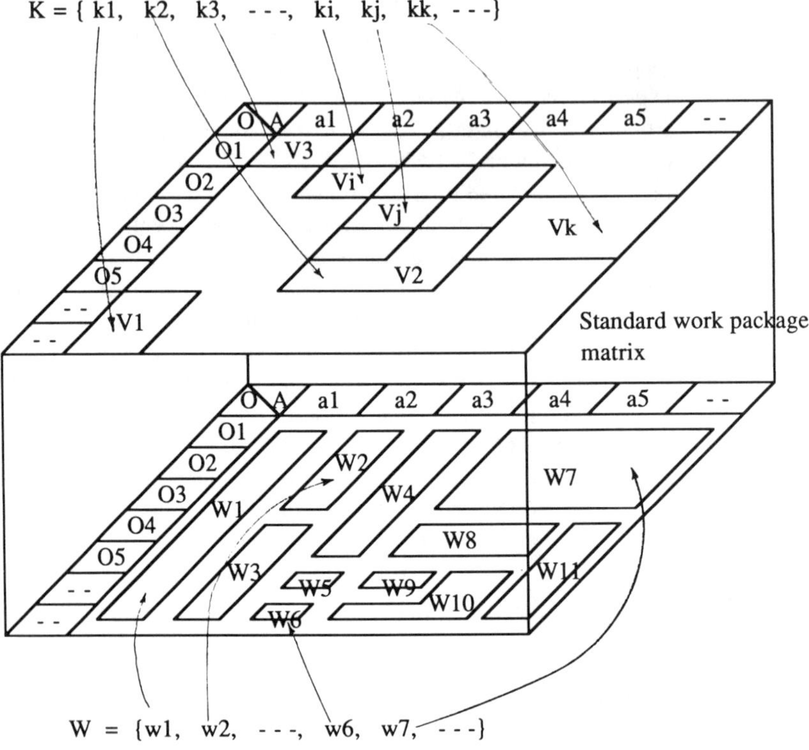

FIGURE 17. Storage of know-how "standard work package matrix" method. (From Niwa, K. and Okuma, M., *IEEE Trans. Eng. Manage.*, 29(4), 146, 1982. With permission.)

the knowledge may be related to the work packages as a suitable index of knowledge.

Figure 18 shows the total framework of the risk management system, and examples of the use of the ES are illustrated in Figure 19. For instance, the project manager may specify a work package and the output data could be risk-reducing strategies that should be followed for that activity. Another example would be to specify a risk as an input and receive as output the risk factors involved together with other possible risks resulting from the original risk factors.

This knowledge-based risk management system for large project execution was developed at the Advanced Research Laboratory, Hitachi Ltd., Japan, on a Hitachi Computer (HITAC M-200), and has been in use for over 7 years, and is the most mature operational ES in the construction industry.

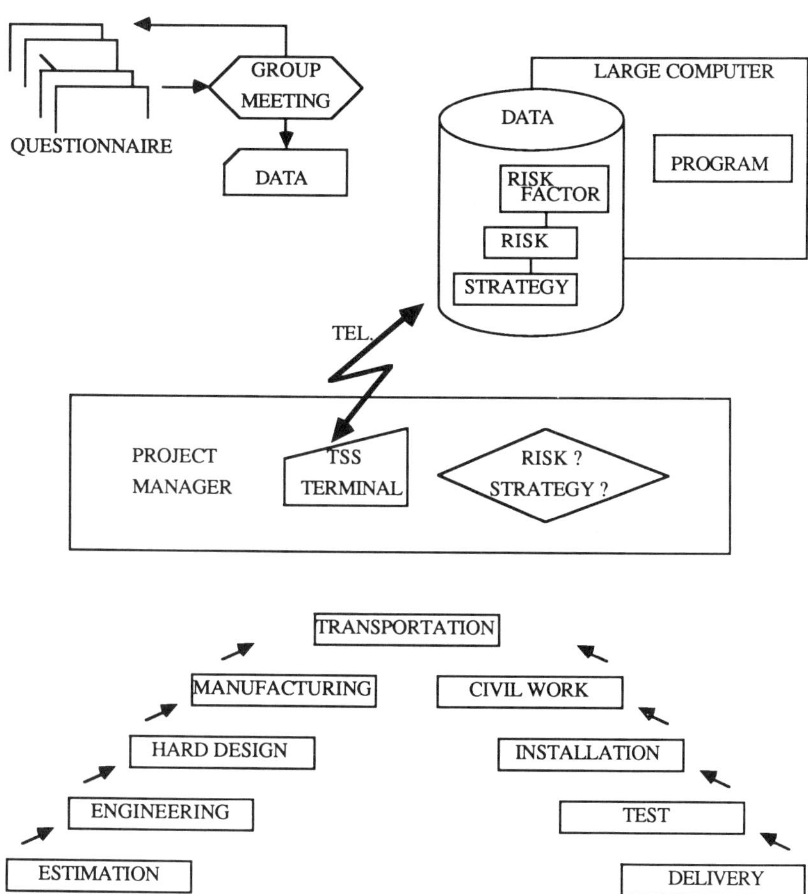

FIGURE 18. Total framework of risk management system. (From Niwa, K. and Okuma, M., *IEEE Trans. Eng. Manage.*, 29(4), 146, 1982. With permission.)

4.3.5. MICROCOMPUTER-BASED EXPERT SYSTEM FOR SAFETY EVALUATION: HOWSAFE (Levitt, 1986)

Stanford's Construction Engineering and Management Program has been involved in construction safety research since 1969. The inadequacy of knowledge dissemination through journal articles and technical reports to jobsite managers motivated the development of HOWSAFE as a convenient means of knowledge transfer to field construction managers.

4.3.5.1. Methodology

HOWSAFE is intended as a diagnostic tool to assist the chief executive of a construction firm in determining the adequacy of the firm's safety programs. It was developed and runs on an IBM PC using The Deciding Factor

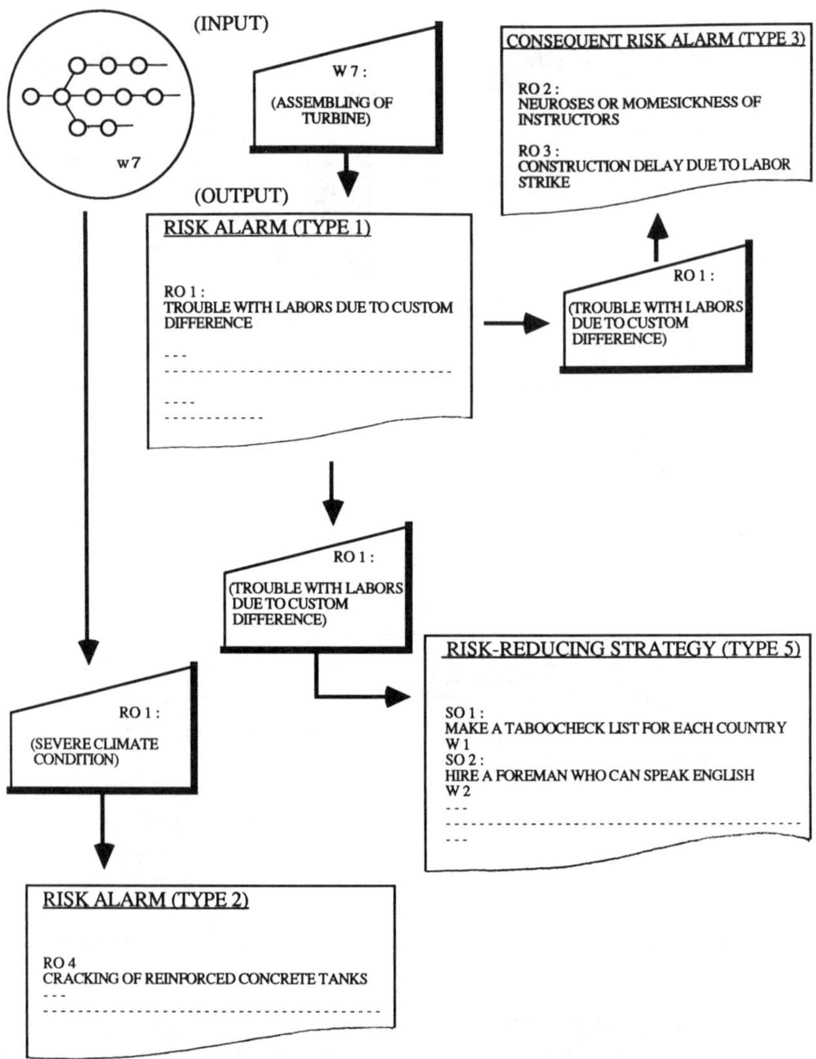

FIGURE 19. Examples of use of know-how transfer method expert system. (From Niwa, K. and Okuma, M., *IEEE Trans. Eng. Manage.*, 29(4), 146, 1982. With permission.)

ES shell, and deals with the diagnosis of an organization's structure and operating procedures. The knowledge to be represented in HOWSAFE starts with a top-level hypothesis: "This construction firm has the required organization and procedures to promote safe construction." A series of intermediate goals such as "Top management truly cares about safety", "Managers at each level are held accountable for the safety of all of their subordinates",

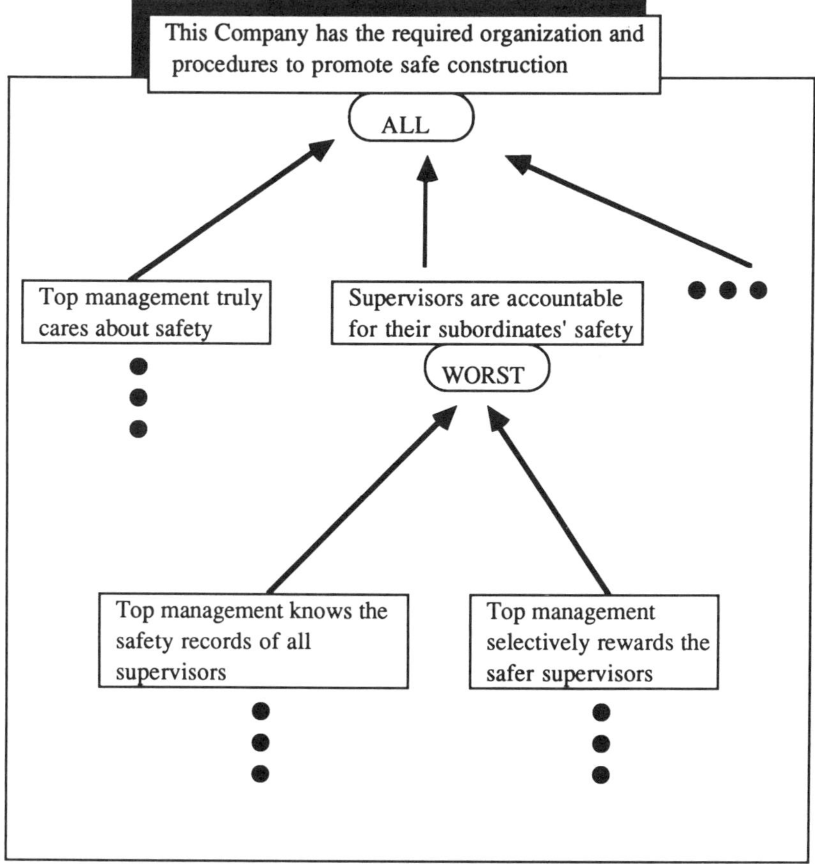

FIGURE 20. A portion of the HOWSAFE inference net. (From Levitt, R. E., Proc. ASCE Symp. Expert Systems in Civil Engineering, Kostem, C. N. and Maher, M. L., Eds., ASCE, Seattle, 1986, 55. With permission.)

etc. lead to the inference of the top-level hypothesis. Each of these intermediate goals is then treated as an hypothesis with lower-level evidence to determine its truth value. The knowledge is structured like an inverted tree, with the top-level diagnosis on the top supported by lower-level inferences whose validity can be evaluated by the user at the bottom end of each branch. This approach to structuring knowledge is essentially equivalent to a production rule system with certainty factors in which the rules are organized hierarchically. Figure 20 shows a portion of the inference net for HOWSAFE.

The Deciding Factor provides the control structure with backward chaining. KILL values and CONDITIONAL logic, which are extensively used in HOWSAFE, permit the system to be tailored such that the user's responses are sought only when needed and consultations have an easy and logical flow. The Deciding Factor has an attractive feature that permits a user to backtrack in a consultation and change a response previously entered. Starting from the top-level hypothesis, the program attempts to satisfy the first goal at the next level. It then chains down, through the first piece of evidence listed at each level, to the bottom or "leaf nodes" of the tree which have no supporting evidence from which their belief can be inferred. This form of knowledge representation was derived from the PROSPECTOR ES developed at SRI in the mid-1970s. A final degree of belief in the top-level hypothesis is reached by combining and weighting the user's responses to leaf node questions.

HOWSAFE has undergone limited external validation and is an operation prototype ES. A comparison package, SAFEQUAL, underwent field testing that resulted in some minor refinements, and is an operational ES.

SAFEQUAL, also developed using the *The Deciding Factor,* helps construction managers to select contractors based upon their past safety performance and current safety management practices.

4.3.6. CONSTRUCTION SCHEDULING KNOWLEDGE REPRESENTATION: CONSAES (O'Connor et al., 1986; DE LA GARZA AND IBBS, 1987; ADELI, 1988)

4.3.6.1. Introduction

Construction scheduling together with estimation, cost control, and quality assurance is an essential ingredient of effective project control. The delivery of a completed facility on time is often more important to a client than cost, especially for revenue-generating projects. One of the primary concerns of today's claims-conscious construction industry is the ability to forecast the likelihood of project disputes and analyze their origins to assign liability. The U.S. Army Corps of Engineers is very interested in the development of an ES that will help Army resident engineers to forecast construction schedule variations, the reasons for those deviations, and the parties responsible. Under a multiyear research contract, the University of Illinois Construction Engineering Expert System Laboratory (CEESL) and the U.S. Army Corps' Construction Engineering Research Laboratory (CERL) are working in collaboration to develop a PC-based ES for analysis of construction schedules.

The development of a knowledge-based ES for construction scheduling is an evolutionary process. The knowledge architecture schemes of semantic net, frames, and object-oriented programming provided significant improvements in the representation of heuristic information. With further progress in research, a general knowledge categorization scheme has been developed to divide scheduling analysis and evaluation into two areas: an initial scheduling analysis module and an in-progress scheduling analysis module. Figure 21 represents the knowledge structure with initial and in-progress scheduling analysis modules based upon the major subcategories of cost, time, logic, and general requirements. The initial schedule analysis module provides the

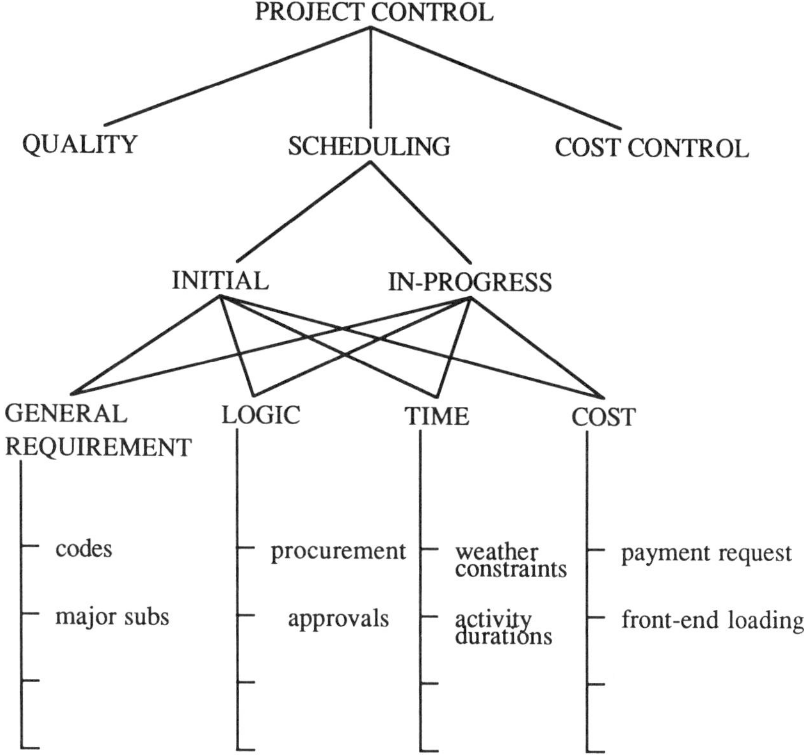

FIGURE 21. Knowledge structure. (From O'Connor, M. J., De La Garza, J. M., and Ibbs, C. W., Jr., Proc. ASCE Symp. Expert Systems in Civil Engineering, Kostem, C. N. and Mayer, M. L., Eds., ASCE, Seattle, 1986, 55. With permission.)

type of information that contractors present to owners for verification before the commencement of the project. Typical information would consist of the owner's approval activities, participation of major subcontractors in the formulation of the plan, etc. The in-progress scheduling evaluation module allows project managers to examine questions such as delay and duration modification concerns.

4.3.6.2. Methodology

CONSAES (CONstruction Scheduling Analysis Expert System) relies upon existing project control system software to (1) identify and capture expressions of similar form in the "paper" knowledge base, (2) determine the specific target inference engine, (3) decide how the "paper" knowledge base is to be represented in the inference engine, and (4) develop a mapping technique to adapt the concepts, facts, and rules to the corresponding engine syntax.

4.3.6.3. Knowledge Organization

As the "paper" knowledge base became larger, it exhibited some regularity (i.e., expressions of similar form frequently reappeared). These regularities were then captured by building an English-like knowledge acquisition grammar. The facts, rules, and concepts of the construction schedule analysis domain are expressed using this grammar. For example, the syntax for the rule and condition categories is:

```
<rule>        :: = IF <conditions> THEN <conclusions>
<condition>   :: = <frame> HAS <parameter> OF <value>
<condition>   :: = <frame> IS IN CLASS <frame>
```

As a specific example, RULE-111 within the Look-Ahead rule group can be represented by the following English and English-like grammars:

"Paper" knowledge base format:
Make projections based on what has happened vs. what was planned.

Knowledge acquisition format:

```
IF      ((?some-activity IS IN CLASS activities) AND
        (?some-activity IS IN CLASS concrete) AND
        (?some-activity HAS status OF finished/in-progress) AND
        (?some-activity HAS assessment of slow-progress) AND
        (concrete HAS lagged OF (>5)))
THEN    ((?activities IS IN CLASS activities) AND
        (?activities IS IN CLASS concrete) AND
        (?activities HAS status OF unfinished) AND
        (set (?activities HAS new-duration OF (*old delay))))
```

Previous job experience with a particular class of work activities is surveyed for a realistic delay factor. If found, that modifier is then related to all subsequent activities in that class to develop a new anticipated schedule duration. Figure 22 shows the evolution of the knowledge formalization and the advantage of utilizing this generic, intermediate knowledge representation language as a gateway.

4.3.6.4. Knowledge Representation

The Automated Reasoning Tool (ART)™ programming environment has been selected as the inference engine to process the knowledge base. It develops "hypothetical worlds" using the technique for generating, representing, and evaluating static/dynamic alternatives.

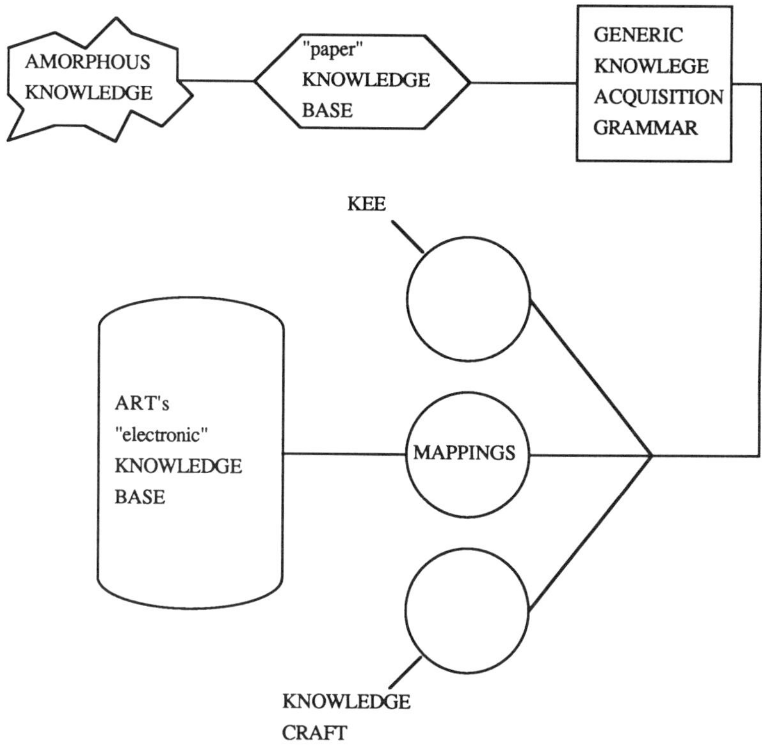

FIGURE 22. Knowledge metamorphosis. (From O'Connor, M. J., De La Garza, J. M., and Ibbs, C. W., Jr., Proc. ASCE Symp. Expert Systems in Civil Engineering, Kostem, C. N. and Mayer, M. L., Eds., ASCE, Seattle, 1986, 55. With permission.)

Object-oriented programming provides the facilities, e.g., objects, to structure information that describes a physical item, a concept, or an activity. Each object is represented as a frame, containing declarative, procedural, and structural information associated with the project. A collection of facts representing an object or class of objects having the same properties constitutes a frame. Using the object-oriented programming feature, ART permits information of a common nature to be stored declaratively in the frames, where it is easily accessible and modifiable.

4.3.6.5. Knowledge Implementation

During the construction planning phase, a work analysis structure is defined based on project phases, goals, and organization. Traditionally, milestone descriptions and codes are defined in such a way that they denote both a building and a construction process, e.g., "cast in place 2nd floor slab". Figure 23 illustrates the hierarchical relationship as well as the inheritance path of a typical milestone. The inclusion of one or more relations in a scheme

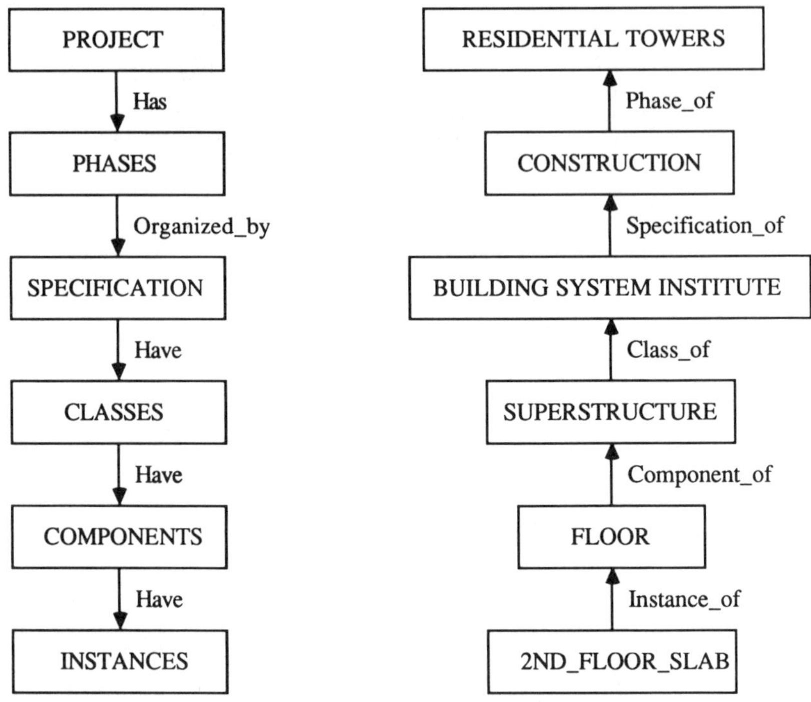

FIGURE 23. Knowledge base taxonomy. (From O'Connor, M. J., De La Garza, J. M., and Ibbs, C. W., Jr., Proc. ASCE Symp. Expert Systems in Civil Engineering, Kostem, C. N. and Mayer, M. L., Eds., ASCE, Seattle, 1986, 55. With permission.)

serves to establish it as a node in a hierarchy. The arrows shown in the diagram have significance in that they originate with the object being defined.

A semantic interpretation of every milestone in the construction schedule is provided by the CONSAES semantic network. For example, when an activity like "cast in place 2nd floor slab" is found in a schedule, CONSAES immediately deduces a series of facts and compilations about it. This activity, for example, contains all basic schedule parameters (e.g., early start, percent complete, etc.), represents a slab in the superstructure, is made of cast in place concrete, consists of formwork, reinforcing steel, and concrete placing, curing, and stripping, is sensitive to cold temperatures, snow, rain, labor productivity, etc.

4.3.6.6. Mappings

A mapping technique adapted to meet ART's specifications relates the English-like knowledge acquisition grammar with ART's knowledge representation language. A different mapping technique needs to be designed for every different inference engine, e.g., ART, KEE, Knowledge Craft (other proprietary, trademarked systems).

4.3.6.7. Relevance of CONSAES

The prototype development demonstrates that this new approach is satisfactory for accelerating and improving the current analyses and computations typical of routine scheduling. CONSAES identifies and organizes the knowledge (analytic and heuristic) useful to the construction engineers to schedule analysis and project management. This ES is ideal for a body of knowledge like scheduling which is partly quantitative and partly subjective.

4.4. EXPERT SYSTEMS IN MAINTENANCE

4.1.1. INTRODUCTION

Mechanical equipment maintenance is a critical function in the operation of plants and facilities. The consequences of component failure or malfunction could be quite serious, and hence evaluations are made regarding the conditions and operating performance of machinery on a periodic time interval. Many variables (pressure, temperature, flow rates, etc.), vibration measurements, and other relevant information are determined, and interpretation of the data requires expertise. Maintenance personnel in the plant are often not experienced enough to interpret the symptoms of problems and determine a remedial course of action. Expert systems (ES) are developed to help the less experienced people to resolve problems with malfunctioning or failed equipment.

4.4.2. CENTRIFUGAL PUMP FAILURE DIAGNOSIS: PUMP PRO™ (Finn and Reinschmidt, 1986)

Reliable operation of pumps at most power and process plants is critical. Correction of pump failures often necessitates the use of expensive and time consuming consultants. Although PUMP PRO™ is principally intended to diagnose pump failures at power and process plants, it can also be used to diagnose pump problems by on-site personnel during the start-up phase. Mechanics, technicians, etc. can avail themselves of expert knowledge in the program without the necessity of calling in a human expert consultant.

4.4.2.1. Methodology

The program is written in MAIDS™, Microcomputer Artificial Intelligence Diagnostic Service, which is a proprietary ES shell developed at Stone and Webster Engineering Corporation (SWEC). The MAIDS inference mechanism is a forward-chaining, rule-based program that uses a subset of the English language for representing rules. The program has two modules, a rule compiler and an execution module.

The operation of the program is separated into four major phases:

1. **Identification of the symptoms**—This is accomplished by means of the MAIDS™ user interface, which consists of text displays and a

Stone and Webster Engineering Corporation
P U M P PRO (w/monitor)
Microcomputer Artificial Intelligence Diagnostic Service

WHICH OF THE FOLLOWNG DESCRIBE THE PUMP CAPACITY

1. PUMP CAPACITY IS ZERO
2. PUMP CAPACITY IS INADEQUATE
3. PUMP CAPACITY IS ADEQUATE

ENTER THE NUMBER CORRESPONDING TO YOUR CHOICE ? 3

FIGURE 24. Typical questions and associated text display in pump failure diagnosis. (From Finn, G. A. and Reinschmidt, K. F., Proc. ASCE Symp. Expert Systems in Civil Engineering, Kostem, C. N. and Mayer, M. L., Eds., Seattle, 1986, 40. With permission.)

question/answer input format. A typical question and associated text display are illustrated in Figure 24.

2. **Identification of the causes**—The program uses its forward-chaining inference procedure to apply heuristic rules to the observed symptoms for identification of the causes.
3. **Provision of tutorials**—The program includes a series of optional tutorials aimed at helping the user understand terminology and procedures. These tutorials are invoked at the user's request, so that users who are familiar with the terminology may proceed directly with the program. In this manner, different levels of user groups can be accommodated without compromising the efficiency or accuracy of the program's operation.
4. **Suggestion of remedies**—After identification of probable causes, the program will instruct the user on appropriate remedial action. If the solution of the problem is beyond the user's capabilities, he will be advised to call in a technical specialist.

PUMP PRO™ diagnoses problems by means of 22 possible symptom classes and a summarized pump history. It allows input of multiple symptoms and provides seven extensive tutorials and many minor tutorials with approximately 350 problem identification rules. A total of approximately 70 rules deal with appropriate remedial strategies and actions. Figure 25 illustrates a sample rule using the MAIDS English-like format extracted from PUMP PRO.

PUMP PRO is a mature operational system, and is one of a family of similar systems offered by SWEC, accessible by modem using an IBM PC-class computer. Users are assessed a charge based on connect time to the SWEC IBM-AT computer in Boston. The present configuration for on-line

```
        BEGIN RULE
CATEGORY          : 16
AUTHOR            : T.J.FRITSCH
DATE              : 3-29-1985
REASON            : EMPIRICAL
CONDITIONS        : PUMPED LIQUID IS CLEAN
ACTIONS           : CLEAR SCREEN
                      DISPLAY BLOCK TEXT
                      'CHECK SHAFT SLEEVES AT PACKING
                      END BLOCK TEXT
                    ASK IS SHAFT/SHAFT SLEEVE WORN
END RULE
```

FIGURE 25. Sample rule using the MAIDS English-like format. (From Finn, G. A. and Reinschmidt, K. F., Proc. ASCE Symp. Expert Systems in Civil Engineering, Kostem, C. N. and Mayer, M. L., Eds., Seattle, 1986, 40. With permission.)

access by users enables the user's PC to act as a terminal to SWEC's IBM PC AT, hosting both the ES shell and the knowledge bases. Communication through the modem makes the program run rather slowly, especially with the large quantity of text that this system must send to the user's screen. The concept of a large consulting firm acting as a dial-up ''knowledge utility'' for many kinds of routine consulting services is unique and challenging.

4.4.3. VIBRATION ANALYSIS INTERPRETATION

The process of diagnosing problems in rotating machinery is dependent, to a large extent, on two factors: (1) the data required in order to make a diagnosis and (2) the expertise of the diagnostician in interpreting the data. Vibration monitoring and measuring is an important art in routine maintenance. Experts in this field can identify causes of vibration after examination of very few typical data. This program was developed at SWEC in order to improve the performance of engineers who are assigned the task of vibration diagnosis.

4.4.3.1. Methodology

The program is an operational ES, which was developed by SWEC using the ES shell EXSYS. It is designed to run on standard, IBM PC-class microcomputers. The inference mechanism uses subroutines for the purpose of analyzing the output of a data collection device and for presenting graphic displays of the analysis results. A VAX-based version has also been implemented, using the inference mechanism installed on the SWEC VAX. The program operates in an interactive question and answer format, and acquires most of its required information from the user or from the output of its own frequency analysis software. The system is rule based, containing over 100 rules, and is able to diagnose 18 separate causes of vibration. The program presents the user with a ranked list of probable causes of vibration and provides fairly detailed explanations of each.

4.4.4. FIELD DIAGNOSIS OF WELDING DEFECTS (Finn and Reinschmidt, 1986)

Welding defects, which are common on most construction sites, can drastically impair construction schedules and escalate project costs. Weld repairs are extremely expensive and in certain cases can have more adverse effects than the defect itself. SWEC has developed an ES to identify the causes of defects and recommend procedures for ensuring welds free from defects. This interactive system allows field personnel, welders, supervisors, or quality control personnel to determine probable causes of weld defects. The program takes into account different welding procedures, code requirements, site conditions, and observations. It enables more rapid repair of welding defects, thus reducing repair costs.

4.4.4.1. Methodology

The system is an operational ES and requires the welding supervisor to answer specific questions about observations made at the site of the weld, the condition of the materials and the environment, and details about the welding procedure employed. The system uses a backward-chaining mechanism to reason about likely causes of the defects. A ranked list of possible factors responsible for the defect is presented to the user together with methods for improving the welding operation.

Parts of the program have been implemented, while other modules are still under development. The weld diagnosis program is written using the ES shell EXSYS for use on an IBM PC-class of microcomputer.

4.4.5. EXPERT SYSTEM FOR CONCRETE PAVEMENT EVALUATION AND REHABILITATION (Hall et al., 1988)

Concrete pavement evaluation and rehabilitation is a complex engineering problem, in view of the large number of interacting factors and the lack of adequate analytical models to solve all aspects of the problem. Successful concrete pavement evaluation and rehabilitation currently relies heavily on the knowledge and experience of authorities in the pavement field for diagnosis of the causes of distress and selection of feasible rehabilitation techniques that cost effectively correct the deterioration. A practical and comprehensive ES has been developed to assist practicing engineers in concrete pavement evaluation and rehabilitation; it uses a new, innovative approach that combines human knowledge and analytical techniques into a user-friendly personal computer program.

4.4.5.1. Methodology

The ES consists of computer programs, one for each of three concrete pavement types: jointed reinforced concrete (JRCP), jointed plain concrete (JPCP), and continuously reinforced concrete (CRCP). The steps in evaluation and rehabilitation design are:

1. Project data collection. The engineer collects key inventory (office) and monitoring (field)data for the project. Inventory data (including design, traffic, materials, soils, and climate) and monitoring data (consisting of distress, drainage characteristics, rideability, and other items collected during a field visit to the project) are entered into a personal computer using a full-screen editor. The overall condition of the project is extrapolated by the system from the sample unit monitoring data.
2. Evaluation of present condition. All the data are analyzed using the evaluation decision trees and major problem areas, including roughness, structural adequacy, drainage, foundation stability, concrete durability, skid resistance, and shoulders, are identified and evaluated; five additional problem areas (transverse and longitudinal joint construction, transverse joint sealant condition, loss of support, load transfer, and joint deterioration) are evaluated in the case of JRC and JPC pavements. Two additional problem areas, longitudinal joint construction and construction joints/terminal treatments, are evaluated in the case of CRC pavements.
3. Prediction of future condition without rehabilitation. The condition of the pavement for 20 years into the future is projected by means of predictive models, the current traffic level, and the anticipated growth rate. Performance prediction is carried out in terms of serviceability and distress types, viz., faulting, cracking, joint deterioration, and failures (punchouts, steel ruptures, and full-depth repairs) for CRCP.
4. Physical testing. The system recommends specific physical tests to verify the evaluation recommendations and provide data needed for rehabilitation design. Recommended types of testing include nondestructive deflection testing, destructive testing (coring and boring), and roughness and friction measurement. Certain types of deficiencies—structural inadequacy, poor rideability, surface friction, drainage conditions, concrete durability (cracking or reactive aggregate distress), foundation movement (due to swelling soil or frost heave), loss of load transfer at joints, loss of slab support, joint deterioration, and evidence of poor joint construction—may justify physical testing.
5. Selection of the main rehabilitation approach. The most appropriate main rehabilitation approach for each traffic lane and shoulder is selected by the engineer, based upon the evaluation results and subsequent interaction with the system. The options consist of reconstruction (including recycling), resurfacing (with concrete or asphalt), or restoration.

A decision tree has been developed for each pavement type to assist the engineer in selecting the most suitable rehabilitation approach. Figure 26 shows the decision tree for JPCP.

6. Development of a detailed rehabilitation strategy. After selection of a suitable rehabilitation approach, the engineer proceeds to develop the detailed rehabilitation alternative for each traffic lane and shoulder by selecting a feasible set of individual rehabilitation techniques to correct the deficiencies present. This may include such items as subdrainage, shoulder repair, full-depth repairs, joint resealing, etc. A set of decision trees has been developed to guide the rehabilitation strategy development process.
7. Prediction of rehabilitation strategy performance. The future performance of the developed rehabilitation strategy is then predicted in terms of key distress types for 20 years into the future based upon assumed traffic growth. The engineer must evaluate the results and determine whether the strategy provides an acceptable life with an optimum cost.
8. Cost analysis of alternatives. The engineer computes the cost for each item in each rehabilitation technique included in the alternative strategy and determines the total and annual costs for the strategy.
9. Selection of the preferred rehabilitation strategy alternative. The engineer considers the life-cycle cost together with constraints that exist for the project (e.g., traffic control, construction time, available funding, etc.) in selecting the preferred alternative. Based upon estimated initial and annual costs, expected life, and performance and various constraints, the user selects the preferred rehabilitation strategy from among the feasible alternatives available.

The shell used was Insight 2+, developed by Level V Research, Inc. Insight 2+ is a production rule-based system shell in which knowledge is expressed in terms of "if-then" rules. The decision trees are incorporated into the Insight 2+ shell by programming each path down each tree (a path being composed of a set of nodes and connecting branches terminating at a conclusion as a single rule). The system has been developed in both manual and computerized form. The programs operate on any IBM-compatible personal computer.

4.5. CONCLUDING REMARKS AND FUTURE TRENDS

The extent and breadth of work already completed, under way, or in the early conceptual stages indicates that many researchers and practitioners in the construction industry consider expert systems (ES) as offering new and potentially valuable capabilities to support decison making in the industry. The software tools available for building ES applications in construction have improved dramatically over the last 5 years. Systems that can run on IBM

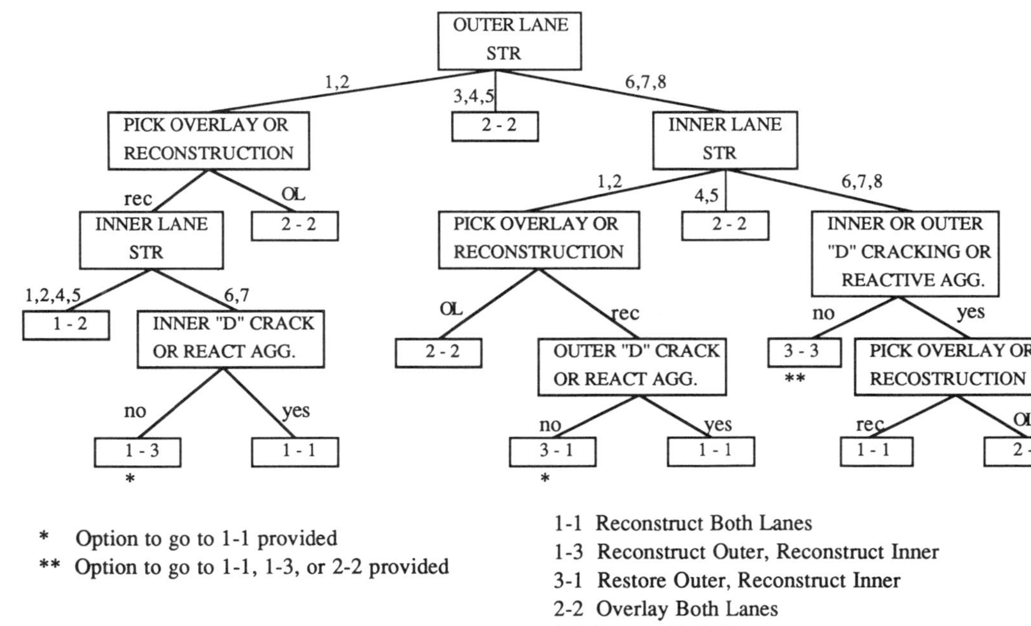

* Option to go to 1-1 provided
** Option to go to 1-1, 1-3, or 2-2 provided

1-1 Reconstruct Both Lanes
1-3 Reconstruct Outer, Reconstruct Inner
3-1 Restore Outer, Reconstruct Inner
2-2 Overlay Both Lanes
3-3 Restore Both Lanes

FIGURE 26. Decision tree for selection of rehabilitation approach for JPCP. (From Hall, K. T. et al., Development of a Demonstration Prototype Expert System for Concrete Pavement Evaluation, Transportation Research record 117, Washington, D.C., 1987, 58.)

PC computers offer outstanding ease of use (The Deciding Factor), the capability to interface with external data and programs (Insight+), and even support of frames (Personal Consultant Plus).

Future research and development of ES in construction will involve hybrid systems combining ES with database management systems and computational systems. The use of ES for integrating between design and construction decision making is likely to be one of the areas for fundamental research and development on ES in construction. ES programming approaches can be used in such hybrid systems to develop individual ES modules, as well as to communicate between these multiple "knowledge sources" and other ES, databases, and application programs. ES in construction can be interfaced with CAD systems, which can attach nongraphical attributes to their graphical objects. Diagnostics for inspection, maintenance, and repair appear to be promising where small ES could be developed for use in desktop or portable personal computers.

Chapter 5

EXPERT SYSTEMS FOR BRIDGE ANALYSIS, RATING, AND MANAGEMENT

5.1. INTRODUCTION

Continuous and systematic maintenance of bridges will extend the service life and reduce the operating costs. Nevertheless, maintenance of bridges and their approaches is often the most neglected phase of highway and railway operations. There are over one half million bridges built on highways and railroads in the U.S. and Canada (PCI, 1975). A modern highway/railroad bridge is a very costly and complex structure, with structural elements from foundation to parapet. The malfunction of one element can affect the overall operational efficiency of the bridge, e.g., pier movement can cause the collapse of an entire span and a damaged bearing seat might cause deck failure. The use of expert systems (ES) has potential for significantly improving design, productivity, and reliability. Bridges are designed to withstand heavy traffic and must be maintained to prevent failures. The costs for the design and periodic maintenance can be reduced with intelligent design, analysis, and management of the bridges (M. Arockiasamy et al., 1993).

5.2. BRIDGE ANALYSIS

The prototype, the Kentucky Bridge Analysis System (KYBAS), is an ES in the area of bridge engineering (Fenske and Fenske, 1990). It is based on a framework of interacting ES for structural analyses and design of highway bridges. This prototype combines a collection of engineering analysis algorithms coded in Fortran with an ES and related interfacing modules in C, resulting in a heterogenous code. KYBAS considers only girder bridges with typical bridge members composed of plate and beam elements. Figure 1 shows the structural elements in a typical bridge. The ES is designed for the following functions:

1. Recommendation of the bridge type, depending upon the location and surroundings
2. Performance of different analyses, including:
   - Static load analyses
   - Dynamic load analyses
   - Cost analyses

The top-level architecture of KYBAS consists of the numeric processing code, knowledge base, and database. The system is developed on a VAX

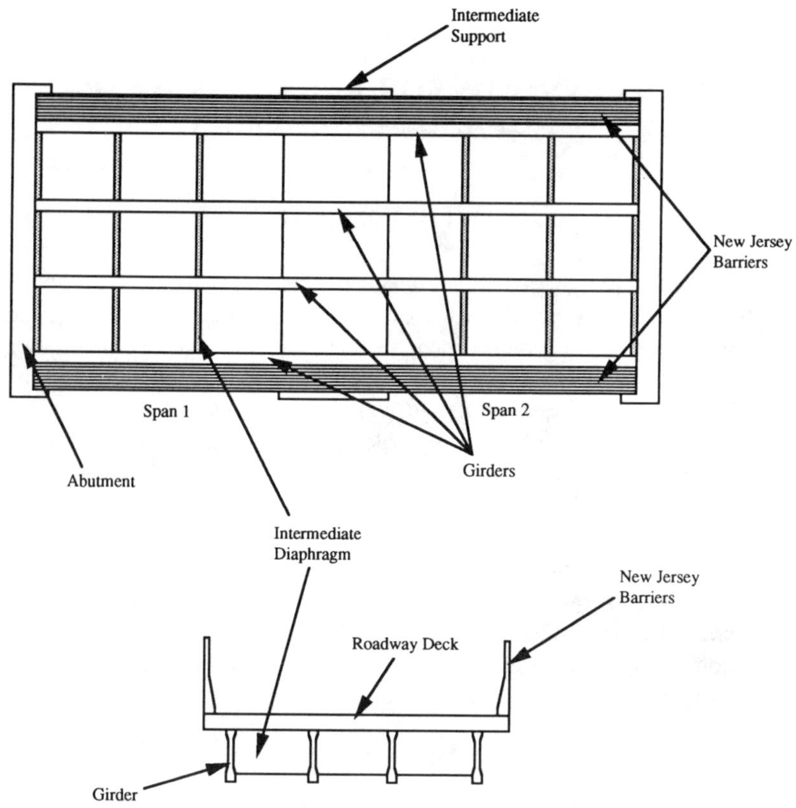

FIGURE 1. Elements in a typical highway girder bridge. (From Fenske, T. E. and Fenske, S. M., *Developments in Short and Medium Span Bridge Engineering '90,* Toronto, August 1990, 23. With permission.)

3200 in the VMS environment. The knowledge base makes decisions about the design parameters of the bridge, including the mesh design and the initial recommendation of the bridge type. The knowledge base component consists of factual knowledge and control knowledge. Factual knowledge represents the current state of the system, whereas control knowledge is responsible for the actions taken during the design decision-making process.

The different groups of rules present in the KYBAS ES consists of superstructure recommendation, SUPERSTR; substructure recommendation, SUBSTR; finite element mesh analysis recommendation, GEOMETRY; an analysis force vector generator recommendation, FORCES; and a preliminary cost estimate, COST. Each group of rules is independent of the other groups so as to allow KYBAS to make recommendations for any bridge component based upon the related expert knowledge. In the SUPERSTR group, KYBAS will recommend the number of spans, bridge width, number of girders,

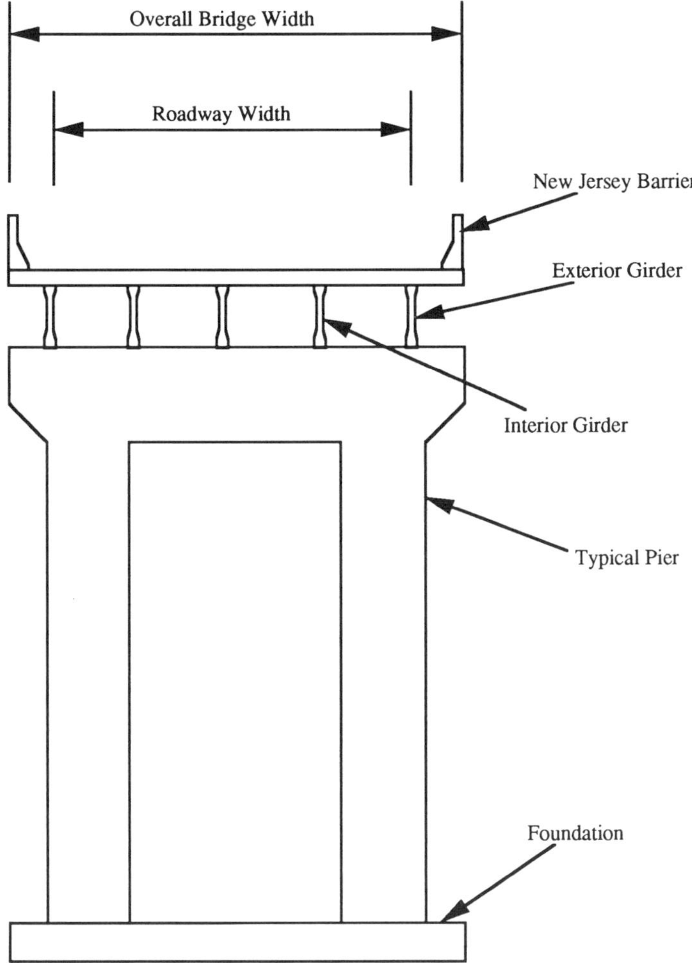

FIGURE 2. Typical bridge configuration recommended by KYBAS. (From Fenske, T. E. and Fenske, S. M., *Developments in Short and Medium Span Bridge Engineering '90,* Toronto, August 1990, 23. With permission.)

AASHTO girder types, diaphragms and their related location in each span, etc. These recommendations are based upon user input to queries regarding clearance, usage, and site location. Figure 2 shows a typical girder bridge configuration recommended by KYBAS.

The prototype RCBD (Reinforced Concrete Bridge Design) ES (Nguyen, 1990) for selecting a reinforced concrete bridge was developed by using the VP-Expert development tool. RCBD is a rule-based ES that has more than 100 rules in its knowledge base. There are 12 different types of bridges for the goal variables and eight dependent variables for the bridge: span length,

TABLE 1
RCBD Knowledge Base Variables

Goal variables	Dependent variables
RC slab bridge	Span length
RC T-beam bridge	(e.g., very short, short, medium long, very long)
RC box girder bridge	Loading
Posttensioned slab bridge	Soil condition
Precast slab bridge	Traffic
Posttensioned girder bridge	Aesthetic
Precast girder and box bridge	Construction
Rigid-frame bridge	Time
Arch bridge	Maintenance
Truss bridge	
Suspension bridge	
Cable-stayed bridge	

From Nguyan, R. P., *Proc. ASEE Annu. Conf.*, ASEE, New York, 1990, 1856. With permission.

TABLE 2
Ranges of Variable Span Length (ft)

Very short	0–49
Short	50–199
Medium	200–599
Long	600–999
Very long	1000 ft or more

From Nguyan, R. P., *Proc. ASEE Annu. Conf.*, ASEE, New York, 1990, 1856. With permission.

loading, soil condition, traffic condition, aesthetic, construction, complete time, and maintenance (Table 1). Table 2 describes the ranges of variable SPAN LENGTH, which varies from very short to very long.

Typical rules in the knowledge base of RCBD are shown below:

ACTION

FIND	Bridge	

RULE 1

IF	Span length	= Short, and
	Loading	= Medium, or
	Loading	= Light, and
	Soil condition	= Normal
THEN	Bridge	= Reinforced concrete T-beams

TABLE 3
Sample of RCBD Expert System Interactive Consultation

kb: RCBD.kbs loaded
Welcome to the world of concrete bridges
RCBD is an expert system to provide advice for bridge selection.

What is the value of SPAN-LENGTH?
Very short
Short
Medium
Long
Very long
What is the value of LOADING?
Light
Medium
Heavy
What is the value of SOIL CONDITION?
Normal
Good
Excellent

The Bridge Selection is Cable-Stayed Bridge

From Nguyan, R. P., *Proc. ASEE Annu. Conf.*, ASEE, New York, 1990, 1856. With permission.

RULE 2

IF	Span length	=	Very long, and
	Loading	=	Heavy, and
	Soil condition	=	Excellent, and
	Aesthetic	=	Very attractive
THEN	Bridge	=	Cable-stayed bridge

A typical run consultation for RCBD is presented in Table 3.

A state-of-the-art integrated bridge design and analysis system (IRDAS) (Needham and Andersen, 1990) has been developed for assessment of the load-carrying capacities of bridges. This system is a fully integrated one which can be used to create three-dimensional finite element analyses of bridge models. The structural assessment is made using the equation.

$$R - L \geq 0 \tag{1}$$

where R is the structure's resistance to a given effect and L is the effect arising from a given loading.

The bridge can be assessed by computing the factored resistance (R_f) of all relevant structural components and comparing them with the factored loadings imposed on the components. The general model structure consists of the following four submodels, each with its own input file:

- Construction model
- Analysis model
- Load model
- Process model

In the construction model input file, the parameters may include the length, width, height, number of beams, materials used, etc. The analysis model file makes it possible to specify a greater degree of detail in the model, if and where desired. The load cases are specified in the load model file for a standard analysis or an analysis of the bridge loaded with an actual vehicle. The control or process file enables specification of the type of construction model, analysis model, and loading cases that should be investigated.

5.3. BRIDGE RATING

The load-carrying capacity and fatigue life need to be determined, and the results for the analyses can be utilized for the design/rehabilitation of a bridge. The computer program BAR6 (PENNDOT, 1989) performs structural analyses of (1) a simple span, reinforced concrete, T-beam bridge or a slab bridge, (2) a simple span, prestressed concrete bridge composed of I-beams, box beams, or plank beams, and (3) a simple or continuous span steel bridge comprised of a deck, stringers, floor beams, and girders or trusses. The girders with in-span hinge and cantilever trusses can also be analyzed using BAR6. The analytical results include reactions, moments, shears, truss member forces, stresses, deflections, rating factors, influence line ordinates for various effects at different sections, etc. All members of the bridge are analyzed and then rated for a set of standard/special live loadings in a single run. The structural and rating analyses performed satisfy the requirements of the AASHTO *Manual for Maintenance Inspection of Bridges* using the working stress method. The fatigue life analysis is carried out in accordance with the Pennsylvania Department of Transportation *Design Manual Part 4*.

The bridge rating system, BRUFEM, is now being developed for bridges in Florida using a finite element model for the analyses (Hays et al., 1990). This consists of three Fortran programs.

1. A preprocessor that develops a finite element model from a relatively small amount of input data about the geometry and stiffness parameters of the bridge
2. A finite element program, SIMPAL, to solve the model created by the preprocessor

3. A postprocessor that uses output from the finite element program and does the bridge rating based on the appropriate service level or strength criteria

The preprocessor and SIMPAL program can be used to prepare and solve finite element models of bridge types, viz., prestressed girders (standard AASHTO), reinforced concrete T-beam girders, structural steel girders, and flat slabs. Edge effects such as parapets can be included or omitted in the models. The loads on the model are generated by specifying the location of the standard or nonstandard vehicles on the bridge. The postprocessor can do the rating of simple or continuous span, prestressed concrete AASHTO girders and simple or continuous span, concrete T-beam girders. Both shear and flexure are considered in the rating of all concrete girder bridges.

5.3.1. RATING FOR CRACKING

The postprocessor calculates the rating factor for prestressed concrete bridges based on a critical tensile stress at the bottom of each girder at girder joint in the longitudinal direction. The rating factor is determined based on the relationship

$$RFC = \frac{\sigma_A - \sigma_{DP}}{\sigma_L} \tag{2}$$

where:

RFC rating factor for cracking
σ_A allowable tensile stress, which varies with $\sqrt{f'c}$ and the type of rating
σ_{DP} stress at bottom of girder due to dead load and prestressing forces
σ_L stress at bottom of girder due to live load

5.3.2. RATING FOR FATIGUE

The postprocessor computes the fatigue rating factor for the concrete T-beam girder bridges based on the range between maximum tensile stress and minimum stress in straight reinforcement steel for the live load cases specified by the user. The rating factor is determined from

$$RFF = \frac{f_A}{f_{rL}} \tag{3}$$

where:

RFF rating factor for fatigue
f_A allowable stress range
f_{rL} range between a maximum tensile stress and minimum stress in straight reinforcement steel due to live load cases.

The allowable stress range is given by the relationship

$$f_A = 23.4 - 0.33\ f_{min} \quad (4)$$

where f_{min} is the algebraic minimum stress level.

5.3.3. RATING FOR ULTIMATE MOMENT

The rating for flexural strength is performed for prestressed concrete and T-beam bridges. Flexural capacity is determined in the postprocessor for each girder based on AASHTO strength criteria, and the rating factor for the ultimate moment is then calculated based on the relationship

$$RFUM = \frac{\phi\ M_n - M_{DP}}{M_L} \quad (5)$$

where:

RFUM	rating factor for ultimate moment
ϕ	strength reduction factor
M_n	nominal flexural strength using AASHTO criteria
M_{DP}	factored girder moment due to all load cases
M_L	total factored girder moment due to live load with impact

5.3.4. RATING FOR ULTIMATE SHEAR

Ultimate shear rating can be performed for the prestressed concrete bridges with standard AASHTO and T-beam girder bridges. The shear capacity for each girder is first determined in the postprocessor based on AASHTO strength criteria. The rating factor for shear for both reinforced concrete T-beams and prestressed girders is calculated by the following formula:

$$RFUS = \frac{\phi\ V_n - V_{DP}}{V_L} \quad (6)$$

where:

RFUS	rating factor for ultimate shear
ϕ	strength reduction factor
V_n	nominal shear strength using AASHTO criteria
V_{DP}	shear force due to all possible dead-load cases
V_L	maximum shear due to live load

The nominal AASHTO shear capacity is calculated for reinforced concrete T-beams. However, for prestressed girders, the shear capacity is given by either flexural-shear capacity or web-shear capacity.

5.4. DATABASE FOR BRIDGE MANAGEMENT SYSTEMS

The database is the foundation of any bridge management system (BMS) (Diaz et al., 1990). The purpose of the database module is to identify all bridges for the BMS, and then be able to review, edit, and/or print related bridge information. The database contains information relative to bridge identification, structure type and material, age and service, geometric data, environment, navigation data, classification, condition, appraisal, load rating and posting, proposed improvements, and inspections. It provides an effective means for quality control of bridge data information. The date base items rated span by span for the bridge include deck condition, superstructure condition, substructure condition, channel condition, culvert condition, structural evaluation, waterway adequacy, critical inspection code and date, scour critical bridge rating, structure length, and deck width.

Chapter 6

EXPERT SYSTEMS IN TRANSPORTATION ENGINEERING

6.1. INTRODUCTION

The practice of transportation engineering has been made more efficent in recent years through computer applications. New software tools such as expert systems (ES) and computer-aided drafting and design (CADD) have been increasingly used on a variety of applications in transportation engineering. Most transportation applications utilize shells for the inference engine, which are proprietary products designed to minimize the amount of computer programming required. The vast majority of transportation-related ES utilize shells specifically written to be compatible with microcomputers.

6.1.1. AUTOMATION FOR TRANSPORTATION (Howell, 1990)

Transportation agencies depend on state-of-the-art automation technology to provide a safe, efficient, and cost-effective highway transportation network for the use of people and communities. The data-processing resources are evenly distributed between engineering and management applications. Figure 1 shows the functional areas in transportation engineering and management under the umbrella of automation.

Automation plays a major role in planning, design, and plan-preparation. The use of satellite surveying technology provides the engineer the ability to establish the position of the proposed construction project, with minimum man-hour effort and high accuracies. After the precise location of the project, engineering maps are prepared using automated photogrammatric methods, which are then distributed through a sophisticated telecommunication system called ETHERNET. The engineer proceeds with the project design and plan preparation using CADD technology. Use of CADD in conjunction with engineering programs such as the Roadway Design System, Integrated Graphics Roadway Design System, Sign Sizing, and Bridge Foundations and Soils Testing enables the engineer to accomplish the design and plan sheets three to four times faster than is possible without the aid of such tools.

The blending of automated engineering and management systems ensures that efficient and functional highways are built and maintained. The management and information system (MIS) incorporates the engineering data into one database, which is then available to support design, construction, finance, materials, equipment, human resources, maintenance systems, and future expansion possibilities. The structure of the MIS is shown in Figure 2.

After the engineer's plans are completed, the MIS is used in generating the design and construction information system (DCIS). DCIS includes the

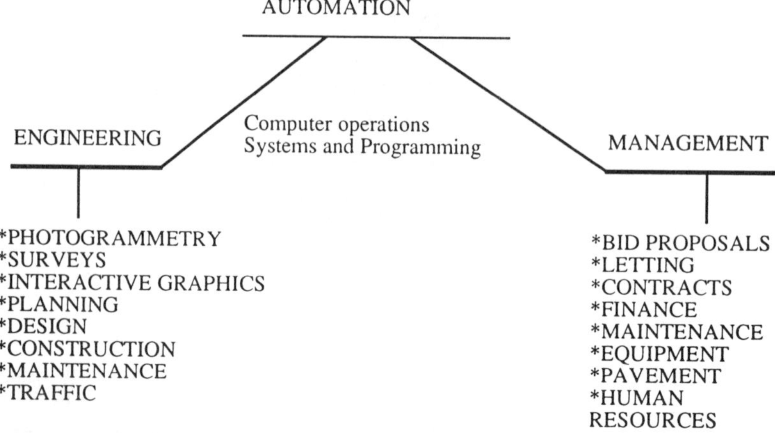

FIGURE 1. Automation umbrella. (From Howell, T. F., *J. Transp. Eng.*, 116(6), 831, 1990. With permission.)

automated bid proposal system, including specifications, general notes, and the plans and letting system that analyzes contractor proposals, determines the successful bidder, and assists the administrator in the contract award. The postletting phase of DCIS consists of an advanced program that receives project data regarding the successful contractor from the letting process and provides assistance to the project manager regarding project oversight, quality assurance, and contractor payments.

After project completion and opening to traffic, roadway information records are maintained regarding accidents, maintenance records, and traffic counts. The maintenance MIS materials supply management system, equipment operating system, and roadway information system assist the department in planning and monthly maintenance activities. All of the above management systems are then integrated into the financial information management system of the MIS. The human resources management system is also an important component, which provides managers with resource information.

The automated engineering systems should ideally work in concert to provide the manager with timely and accurate information, thereby providing valuable information to the decision-making process.

6.2. INTEGRATED TRAFFIC DATA SYSTEM (ITDS) (Rathi et al., 1990)

Traffic simulation, signal timing, and optimization programs or models have been developed by the Federal Highway Authority and others over the past 20 years. Extensive use of these computer models has demonstrated their potential as effective tools in the development and evaluation of traffic control

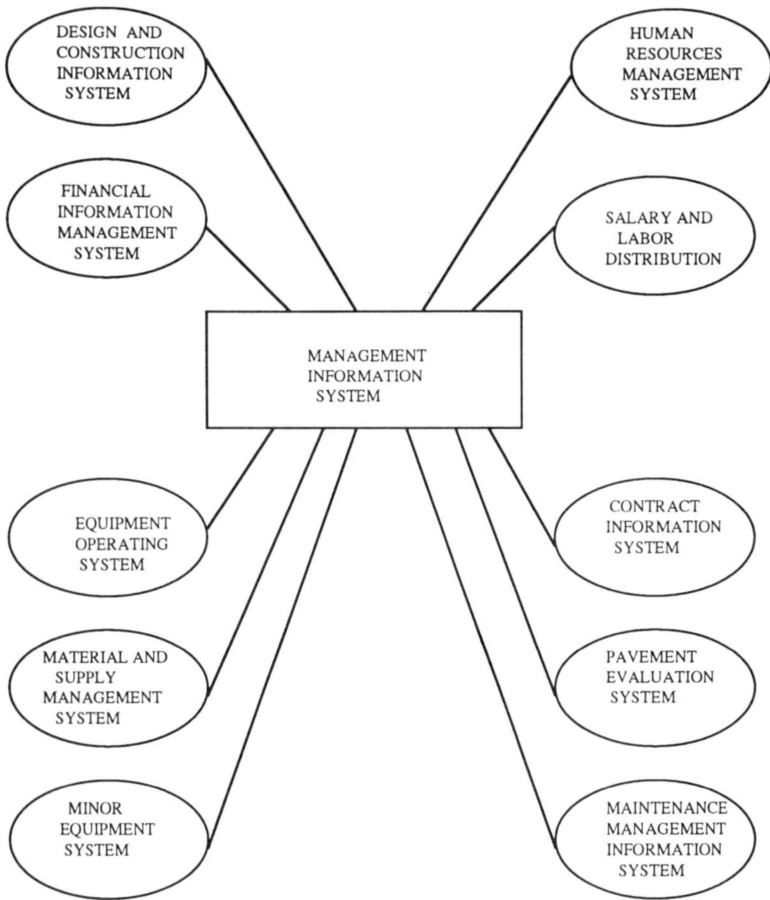

FIGURE 2. Management information system. (From Howell, T. F., *J. Transp. Eng.*, 116(6), 831, 1990. With permission.)

strategies. Transportation engineers can design and evaluate various alternative traffic control strategies prior to their real-life implementation. This evaluation helps reduce motorists' operating costs, vehicular fuel consumption and emissions, and costly retrofits, which occur when a problem is detected only after implementation.

Traffic models, however, have a few major difficulties associated with their use. Basically, there is no consistency in the definition of traffic-related terminology used by different models. Also, since these computer programs were written for use in a batch processing mode, a card image input data file must be prepared for each run. This makes them awkward and time consuming to use. Finally, due to the limitations of these models, users who are developing comprehensive traffic control strategies are required to use more than one model, so one who uses several of these models must become familiar with the data and input format requirements of each.

ITDS was sponsored by the Federal Highway Administration in 1982, as a response to these and other deficiencies. It provides an easy to use "front end" to some of the most commonly used traffic models. ITDS creates data files that can be used directly as input to the various models supported by it.

ITDS is a microcomputer-based system through which transportation engineers can store, maintain, and update traffic network information in a centralized database. This information is then used to create input data files of some of the common traffic simulation and network signal timing optimization models. Thus, ITDS provides a friendly front end or preprocessor, by using the concept of an integrated traffic database. This database provides a means of storing, maintaining, and retrieving traffic data in a format that is independent of any traffic model, and hence the database can be used with little or no modification for other traffic models that may be supported by ITDS.

In order to maintain a database effectively and systematically, a specialized application programming software called a database management system (DBMS) should be used. The ITDS incorporates a commercial network DBMS called MDBS-III. All the details of storing and retrieving data from the database are handled by the MDBS-III software. Figure 3 shows the structure of the network database used by MDBS-III. This type of structure allows the user to store and maintain all information relevant to a traffic network in a generic database, without making much modification.

As seen in Figure 3, the various logical groupings of data concerning the characteristics of the traffic network are detailed. These logical groupings are linked together by the use of set definitions. A set consists of a relational ownership condition. For example, NETWORK, LINK, and NODE are defined as sets. Each of these sets contain one or more of these attributes (e.g., the set LINK contains 28 attributes). The set LINK is defined to be part of set NETWORK and also attached to several other sets such as volume and LANE, and this relationship between various sets is detailed.

The structure of ITDS is shown in Figure 4. Since the structure of ITDS was designed with future expansion and modification possibilities in mind, the design is highly modular in nature. New and revised models can be interfaced to the ITDS simply by adding new model interfaces, since the database structure incorporated in the system is independent of any particular model's requirements.

The ITDS consists of four major components: screen unit, database editors, database access programs, and traffic model interfaces ("card deck formatters").

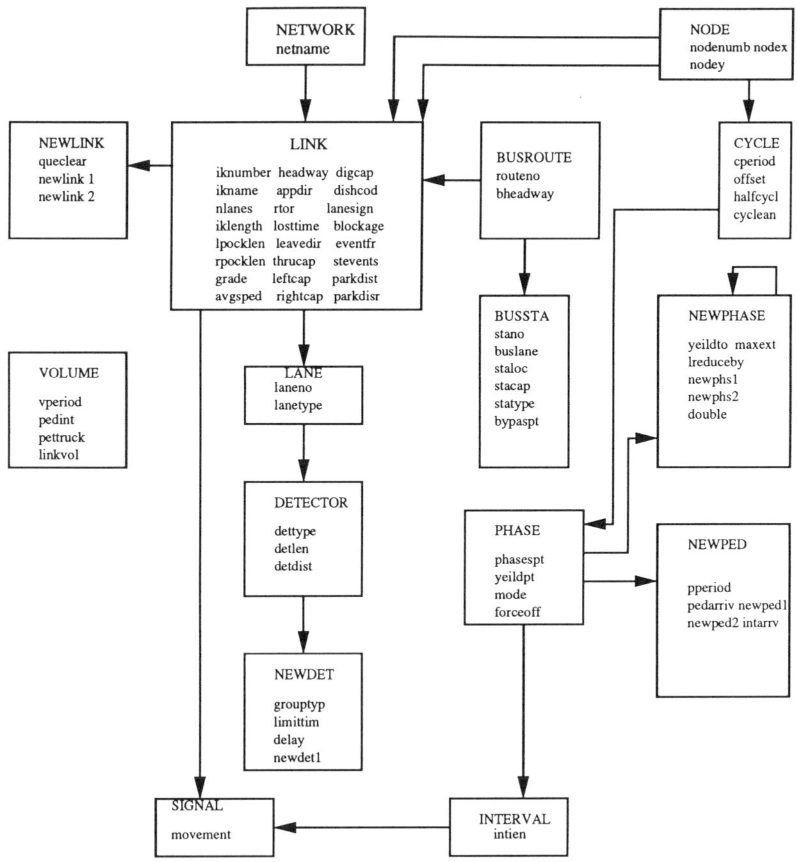

FIGURE 3. ITDS database scheme. (From Rathi, A. K. et al., *J. Transp. Eng.*, 116(6), 799, 1990. With permission.)

6.2.1. SCREEN UNIT

The interaction between the user's terminal and all other ITDS component programs is handled by this component. Basically, the screen unit component consists of a set of routines that are called from various program components to display text and default values on the screen, accept user responses, and help the user to understand the screen fields and requirements. Routines are present that also perform some elementary input range checking for the input values entered by the user.

6.2.2. DATABASE EDITORS

This component of the ITDS consists of a set of programs that provide the user with the capability to display, add, delete, and modify information

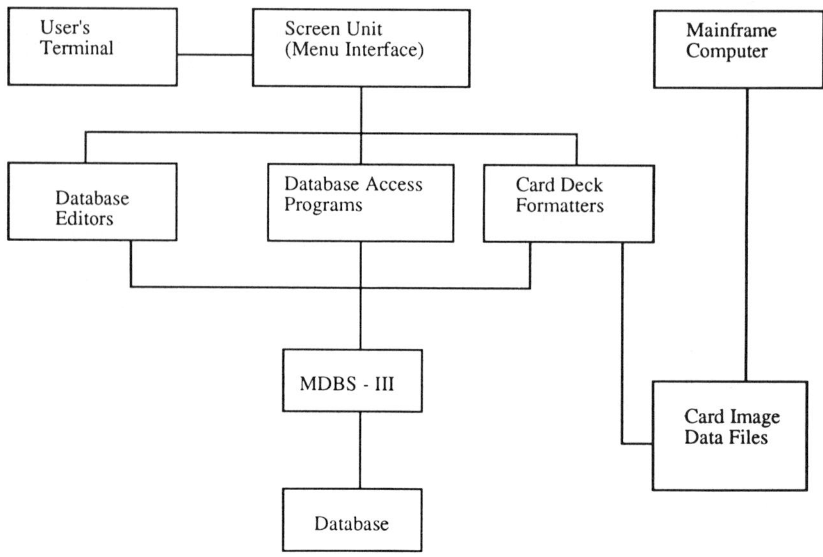

FIGURE 4. Structure of ITDS. (From Rathi, A. K. et al., *J. Transp. Eng.*, 116(6), 799, 1990. With permission.)

stored in the traffic database, either alphanumerically or graphically. These database editors provide the user with an easy-to-use interface with the database.

6.2.3. DATABASE ACCESS PROGRAMS

These programs enable the user to output from the database. Functions such as print, query or graphically displaying the information contained in the database are achieved through these programs.

6.2.4. CARD DECK FORMATTERS

Through this component of ITDS data from the database are extracted to produce the actual card image files that serve as input to the traffic models.

In addition to these components, two new features were added to the ITDS that greatly enhanced the usefulness of the system: The interactive graphics data input and the static graphics display.

6.2.5. INTERACTIVE GRAPHICS DATA INPUT

In the original version of ITDS, the user was required to input data in the form of numbers and codes. This representation of data can cause the user to lose the intuitive sense of the relationship of traffic-related information to the actual network geometry. The traffic engineer had to refer back and forth to street maps, network diagrams, and the traffic model's manuals to verify the accuracy and compatibility of the data. By incorporating the interactive graphics data input feature, the user can create and modify the basic components of a traffic network representation by effectively drawing a link-node diagram on the screen.

With the graphics input facility, a node can be added at a certain location by simply pointing the cursor to the corresponding spot on the computer's cathode ray tube (CRT) screen, rather than entering the coordinates of that point through the screen unit, as in the previous version, by filling fields with numbers and alphanumeric codes. Modifying the database in this way is less prone to error, since the user can see the changes on a map of the actual traffic network.

6.2.6. STATIC GRAPHICS DISPLAY

The graphics display facility allows the user to see a traffic network graphically depicted on a device such as the CRT screen or a plotter. Four kinds of graphics displays are provided. The first is the network link-node diagram, on which the various attributes of links can be displayed by printing number values alongside the links or by a graduated shading of the links. The second type of graphics display is the network lane-detailed diagram. Here, a more detailed diagram of a network can be created by using information about lanes and turning pockets along with node locations and other link information.

Another kind of graphic display is a lane-detailed intersection diagram. Here, an enlarged view of a single intersection can be depicted, along with a small link-node diagram of the intersection and a listing of the values or chosen attributes for the links entering the intersection. The last type of graphic display is a detailed link diagram, showing many attributes of a link and its lanes. Icons are used to represent associated items such as detectors and bus stations.

Currently, ITDS runs on an IBM or compatible PC under MS DOS 3.0 or higher versions. It provides interfaces to TRAF-NETSIM version 2.00, TRANSYT-7F release 5.00, and PASSER II-84 version 3.00. The ITDS, in conjunction with the microcomputer versions of the various available traffic models, will provide engineers and analysts a powerful analytical tool for developing and testing alternative traffic control strategies.

6.3. KNOWLEDGE-BASED EXPERT SYSTEM FOR MANAGING INTERSECTION SAFETY (Senerviratne, 1990)

In recent years, detailed studies of the causes of traffic accidents and the means of alleviating road or traffic hazards have been carried out by agencies and researchers. The results of these studies have not been available in an accessible form to all agencies responsible for managing traffic safety. Due to budgetary or other resource constraints, many agencies responsible for traffic safety have not been able to utilize the knowledge of safety management techniques. Instead, they must rely on small knowledge bases founded mostly on local experience and handbooks or procedural manuals. With properly designed knowledge-based systems, the experience and knowledge based on the outcome of years of research talent and millions of dollars of research budgets can be made available to all. A knowledge-based system for managing traffic safety at intersections is discussed here. It is designed to assist the engineer to establish possible causes of accidents and to identify countermeasures. The option to evaluate the cost effectiveness of each countermeasure is also provided.

The Intersection Safety Management Information System (ISMIS) is a menu-driven, microcomputer-based system that provides users with explanations for certain decisions and measures available for the remedy of a particular problem. It also provides an option with which the user can evaluate the cost effectiveness of each remedial measure, and thus determine the best alternative. ISMIS is capable of resolving traffic safety problems specific to four-way intersections. Two basic functions, ''fault detection'' and ''fault diagnosis'', are carried out by a process of comparing the observed conditions at the problem site with idealized conditions. If all the observed conditions do not match with the idealized conditions, further examination is carried out to determine which of the unmatched conditions are significant. Once a site has been identified as being a problem site, the problem has to be defined in terms of three basic types of accidents: (1) left turn, (2) right turn, and (3) pedestrian. Based on the type of problem, the user is required to pick a few answer choices from questions relating to either the vehicle or persons found responsible for the accident (primary units) or vehicles/persons hit by the primary unit (secondary unit). For example, if the problem involves right-turning accidents, the following questions are posed to the user, with the appropriate response given in a multiple-choice format. Then, the most likely causes and the appropriate countermeasures can be determined (Senerviratne, 1990).

- Type of traffic control device (i.e., light, yield, stop signs)
- Direction of travel of secondary vehicle with respect to primary vehicles
- If signal controlled, the phases, cycle time and split
- Approach speeds in each leg, turning movements
- Lighting condition
- Road surface condition

Thus, the ISMIS resembles a standard accident analysis procedure such as the one suggested in the Highway Safety Engineering Studies procedural guide. The difference of a knowledge-based system from a manual approach is that inferences or conclusions at each stage of the analysis are made by a group of people experienced in the field, and the computational speed is thousands of times faster than human abilities. Thus, the quantity of information about a problem and the number of problems that can be analyzed result in a considerable saving of time and man-power. After problem diagnosis is complete, the system suggests various countermeasures along with detailed information like capital and operating costs, implementation procedures, likelihood of effectiveness and sources of reference for each one. There is the option of performing a cost-effectiveness analysis on all or a selected set of the countermeasures suggested by the system.

ISMIS consists of four main components: the database; the knowledge base; the inference engine; algorithmic models for the computation of sight distance, conditional probabilities, and cost effectiveness countermeasures. Figures 5 (a) through (g) give the sample output of a typical session with ISMIS. The first screen enables the user to start the system (Figure 5a). The data input system takes in a keyword, compares it with the descriptors of a typical problem site in the knowledge base, and displays a question and the set of possible answers on the screen, as shown in Figure 5b. The user's input is waited for and decoded, and this information is transferred back to the inference engine, which processes it and comes up with the next condition, the direction of travel (Figure 5c). This process is repeated until the inference engine has enough information to match all the observed conditions with the idealized conditions. If all the conditions cannot be matched, the inference engine selects a suitable match for the unmatched conditions with the countermeasure characteristics. The system provides the flexibility to backtrack to any previous level and redo the analysis, if the user so wishes.

6.3.1. KNOWLEDGE BASE OF ISMIS

The knowledge base contains the information regarding each of the previously defined safety problems at four-way intersections, their countermeasures, and the rules that help determine the causes of each problem. The deviation of a particular observed condition from the corresponding ideal condition will influence the safety. In some cases where the information cannot be precisely defined in terms of numbers, the user needs to make certain assumptions that may or may not be completely true. The user has to respond to a certain question by describing the condition as "good", "average", or "poor". The system needs to verify the contribution of the condition to the problem in more detail. There are many suggestions as to the best possible approach for handling uncertainty and linguistic terms. ISMIS uses a basic proability rule based on Baye's theorem to determine the most significant contributor to a problem:

$$P(C/A) = \frac{P(A/C)P(C)}{P(A/C)P(C) + P(A/\overline{C})P(\overline{C})} \qquad (1)$$

ISMIS: INTERSECTION SAFETY MANAGEMENT INFORMATION SYSTEM

MAIN MENU

START ISMIS
LOAD ALL DATA BASES
SAVE INPUT DATA
COST EFFECTIVENESS ANALYSIS
QUERRY
HELP
DOS COMMANDS

ESC TO RETURN TO PREVIOUS LEVEL OR USE ARROW KEYS TO SELECT AND PRESS RETURN TO ACTIVATE CHOICE

(a)

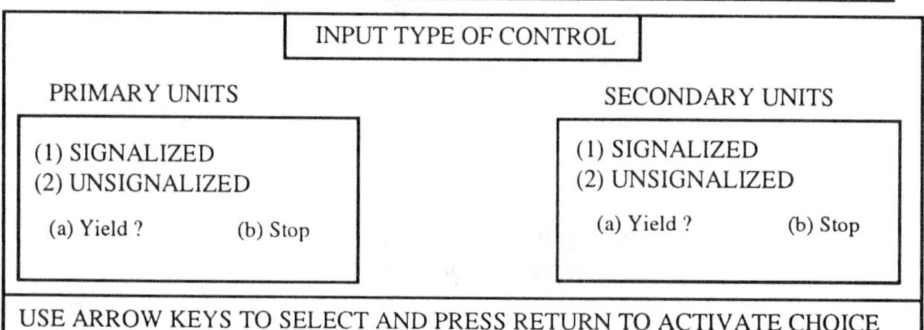

(b)

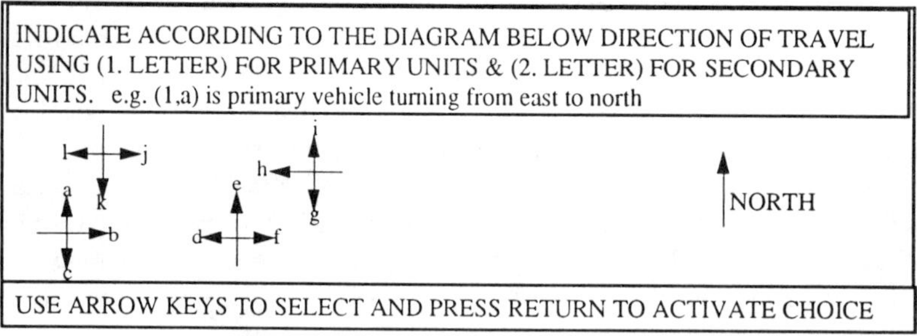

(c)

FIGURE 5. Illustration of ISMIS menus. (From Senerviratne, P. N., *J. Transp. Eng.*, 116(6), 770, 1990. With permission.)

ISMIS: INTERSECTION SAFETY MANAGEMENT INFORMATION SYSTEM

INPUT NAME OF INTERSECTION DETAILS
FILE NAME OR WORK SHEET?......

ENTER FILE NAME AND PRESS RETURN TO ACTIVATE CHOICE

(d)

ISMIS: INTERSECTION SAFETY MANAGEMENT INFORMATION SYSTEM

SELECT THE CONDITIONS OF (1) DELINEATION AND PAVEMENT MARKINGS (2) PAVEMENT SURFACE, & (3) LIGHTING AND ATTACH YOUR DEGREE OF BELIEF IN THE FOLLOWING FORM

(a) good (b) average (c) poor SELECTION : (1, ?, ?); (2, ?, ?); (3, ?, ?)

USE ARROWS KEYS TO SELECT AND PRESS RETURN TO ACTIVATE

(e)

ISMIS: INTERSECTION SAFETY MANAGEMENT INFORMATION SYSTEM

ISMIS. SUMMARY

1) Intersection is signalized
2) Primary units turning east from south are in collision with secondary units travelling west-east.
3) According to approch speeds clearence time is ..% less than required.
4) Approch speeds of primary vehicles exceed maximum permitted according to sight distance by .....%
5) Pavement surface condition is average and degree of belief in condition is .....
6) Lighting condition is average and degree of belief in condition is ....
7) Delineation & Markings condition is average and degree of belief in condition is ....

PRESS RETURN TO VIEW COUNTERMEASURES OR UP ESC. TO RETURN TO PREVIOUS LEVEL

(f)

ISMIS: INTERSECTION SAFETY MANAGEMENT INFORMATION SYSTEM

SUGGESTED COUNTERMEASURES:

1) Reduce speed limit for secondary vehicles.
2) Prohibit right turns on red and post new warning signs to that effect.
3) Prohibit parking on east-west within 10m of intersection to increase visibility
4) Improve delineation to ensure right turns are made into outside lane.
5) Add an all red phase or increase yellow time for primary units.

PRESS RETURN TO MAIN MENU

(g)

FIGURE 5 (continued).

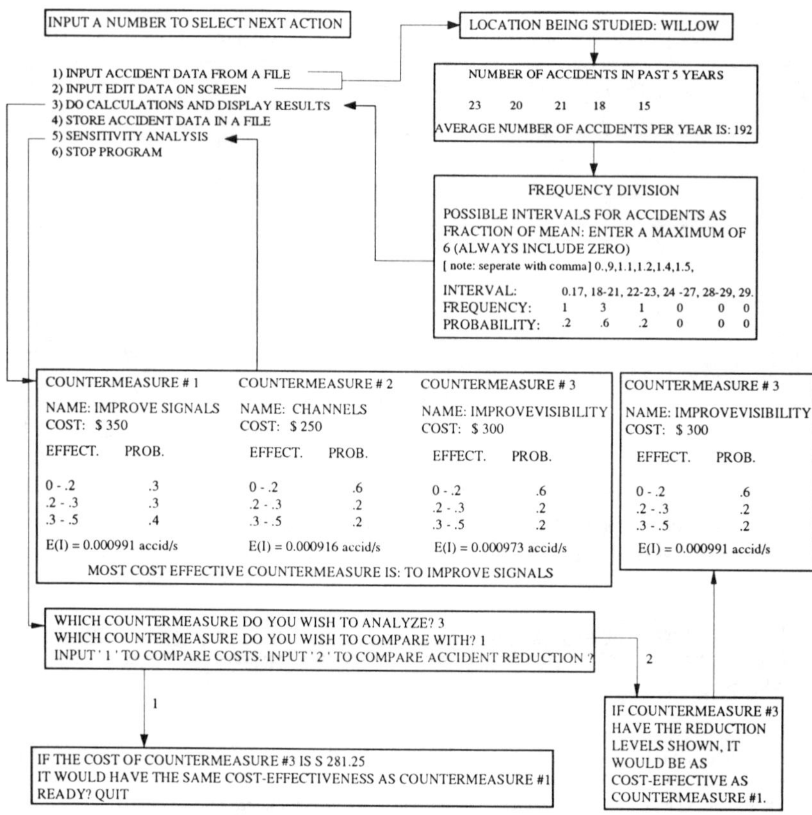

FIGURE 6. Components of countermeasure-evaluation program. (From Senerviratne, P. N., *J. Transp. Eng.*, 116(6), 770, 1990. With permission.)

where P(C/A) is the probability of condition C present, given that a problem exists, P(C) is the probability of the unusal condition being present, and P(A/C) is the probability of a problem existing when condition C is present.

The countermeasures, prioritized according to P(C/A), are stored in the database, and are transferred to the countermeasure evaluation model if the user wishes to perform a cost-effectiveness analysis. Figure 6 shows the components of the countermeasure evaluation program (Senerviratne, 1990).

Thus, the analysis of safety issues can be speeded up through the use of the expert system ISMIS. The most critical conditions can be isolated and the most efficient countermeasure selected.

6.4. AN EXPERT SYSTEM FOR TRAFFIC CONTROL IN WORK ZONES (Faghri and Demetsky, 1990)

Various strategies for the control of traffic in and around work zones have been documented for different problems. It is essential that applications to specific situations are collected and recommended strategies established. Use of the expert systems (ES) methodology for the problem of routing traffic through work zones offers several advantages: (1) the engineer can make fast and accurate judgments or decisions for a task, (2) human expertise in a field is only localized, but the ES can be used by many persons at different locations, and (3) experts are available to develop the system's knowledge base. The objectives for traffic control in work zones involving maintenence or construction works are to (1) protect the roadway user and the work force, (2) have smooth movement of maximum traffic volume (i.e., minimize delays), and (3) achieve efficiency and economy in the work procedures. The knowledge base of the ES is usually created from a literature survey on existing techniques and guidelines.

The development of an ES (called TRANZ) for selecting appropriate traffic control strategies and management techniques around highway work zones is discussed here (Faghri and Demetsky, 1990). The following steps outline the development of the ES TRANZ:

1. Problem identification
2. Conceptualization
3. Formalization
4. Implementation
5. Testing

6.4.1. PROBLEM IDENTIFICATION

Several important dimensions of a problem must be identified and characterized before presenting it to the ES. Information such as the nature of the construction project, the volume of traffic moving on the roadway, characteristics of the road, and the anticipated time the project would take is required. The system would also ask for information regarding the type of roadway (limited access, primary, unlimited access, or secondary) and the position of the lane coming under the work zone area, i.e., inner, center, or outer lane, as shown in Figure 7.

6.4.2. CONCEPTUALIZATION

After the problem has been identified in the first stage, the system recommends a set of solutions. It also provides their relative ratings with regard to potential of utilization and application. The following stages are involved in the problem-solution process:

1. Acquire all necessary information about a project from the user.
2. Answer the relevant questions that the user asks.
3. Determine the relationships between the variables in the system.
4. Provide a list of strategies or devices that are appropriate for each work zone project.

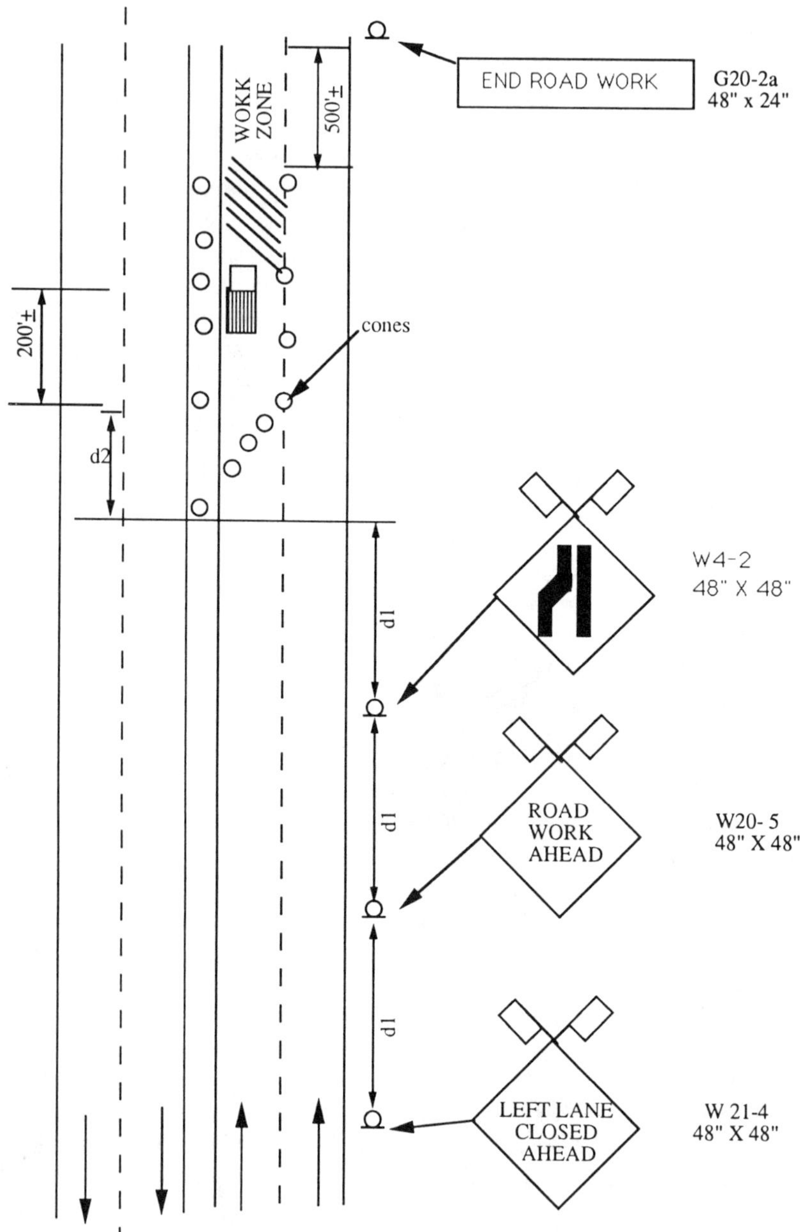

FIGURE 7. Typical maintenance operation on secondary highway. (From Faghri, A. and Demetsky, M. J., *J. Transp. Eng.*, 116(6), 759, 1990. With permission.)

6.4.3. FORMALIZATION

In the formalization stage, the key concepts and relations encountered in the first two stages are expressed.

6.4.4. IMPLEMENTATION

The formalized knowledge acquired in the previous stages is developed into a working computer program in this stage. The following were strategies taken during the programming of the ES TRANZ. The possible problem solutions presented to the user are called choices. The choices in TRANZ include 61 different traffic control devices and management strategies for highway work zones.

Examples for the choices provided by TRANZ are (Faghri and Demetsky, 1990):

1. Road work ahead [W21-4, 48 in. × 48 in. (122 cm × 122 cm)].
2. Road work 1 mi (1.6 km) [W21-4, 48 in. × 48 in. (122 cm × 122 cm)].
3. Trucks must use right lane [VR-26, 48 in. × 60 in. (122 cm × 152 cm)].
4. Next 2 mi (3.2 km) [VR-27, 48 in. × 12 in. (122 cm × 30 cm) under VR-26].
5. Right shoulder closed ahead [VW-28, 48 in. × 48 in. (122 cm × 122 cm)].
6. Right lane closed 3/4 mi (1.2 km) [W20-5, 48 in. × 48 in. (122 cm × 122 cm)].
7. Right lane closed ahead [W20-5, 48 in. × 48 in. (122 cm × 122 cm)].
8. Left lane closed 3/4 mi (1.2 km) [W20-5, 48 in. × 48 in. (122 cm × 122 cm)].
9. Left lane closed ahead [W20-5, 48 in. × 48 in. (122 cm × 122 cm)].
10. Left two lanes closed 3/4 mi (1.2 km) [W20-5, 48 in. × 48 in. (122 cm × 122 cm)].

The rules in TRANZ are in the form of sentences. These sentences are often made of qualifiers and values. The first part of the sentence is the qualifier, usually a verb, and the values are the possible completions to the sentences. Examples of the qualifiers are (Faghri and Demetsky):

1. The nonremovable fixed object(s) existing near the travelway is (are):

    - Exposed work crew
    - Any object considered dangerous to a moving vehicle, such as:

        - Headwall
        - Parapet
        - Manhole
        - Guardrail end
        - Drop inlet
        - Barrier ends
        - Pipe
        - Stored material
        - Bridge pier
        - Equipment
        - Slope
        - Signpoles/bases
        - Box culvert

- An excavation with a depth of 6 in. (15 cm) to 2 ft (61 cm) on or near the travelway
- None of the above

2. The object considered can be:

- Nonremovable fixed object
- Removable object
- Information not available or not applicable

3. The removable fixed object is considered as:

- Hazardous
- Not hazardous
- Information not available or not applicable

All of the variables and parameters described in the formalization process, along with the choices, qualifiers, and values, are now used to form a series of complex inference chains. The rules are created using these inference chains. First, the system determines whether an obstruction near the work zone or project is considered a fixed or nonfixed object. This is done by requesting information regarding the nature of the obstruction and all the characteristics of the obstruction. The system then determines whether the fixed object is considered a removable or a nonremovable fixed object. The user is queried regarding the speed and volume of traffic on the roadway. At this level, the system can decide whether the situations described are hazardous or nonhazardous. Based on this decision, different groups of signaling devices or other traffic control devices and strategies are selected. Finally, the system requests any other relevant information regarding the nature of the project, the different variables associated with the roadway, etc., so that the set of traffic control devices or the traffic management strategies can be further limited to a few relevant options, and then recommended to the user. Figures 8 and 9 show two sample inference chains in the knowledge base and the rules they represent. The complete design process, showing the decision framework in TRANZ, is illustrated in Figure 10.

6.4.5. EVALUATION AND TESTING

After developing the computer program for TRANZ, the system was evaluated. In the evaluation stage, the program was reexamined with a view to improve its efficiency. Logical errors in the program were corrected and the user interface enhanced. The experts who assisted in developing the system were revisited for their comments concerning the prototype and appropriate revisions, if any, carried out. The resultant enhanced prototype was selectively

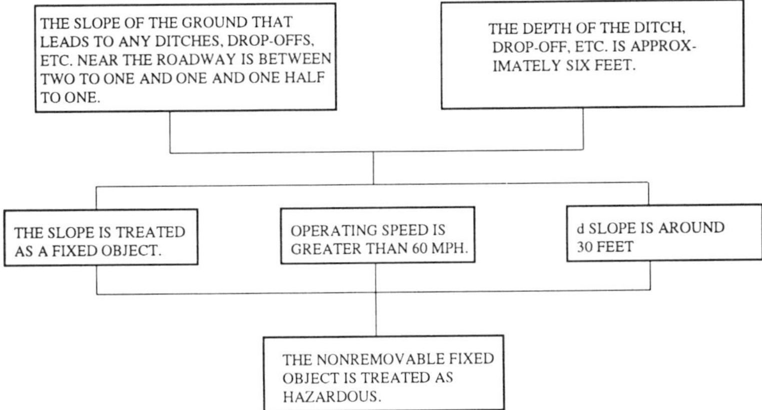

FIGURE 8. Example of inference rule chain in TRANZ. (From Faghri, A. and Demetsky, M. J., *J. Transp. Eng.*, 116(6), 759, 1990. With permission.)

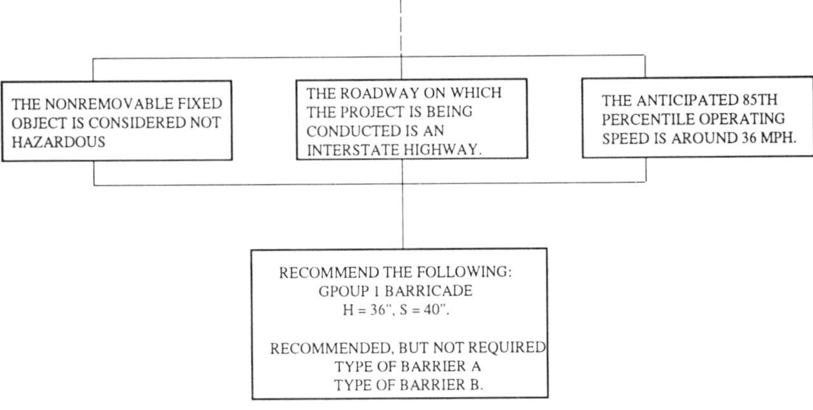

FIGURE 9. Example of inference chain in TRANZ. (From Faghri, A. and Demetsky, M. J., *J. Transp. Eng.*, 116(6), 759, 1990. With permission.)

distributed to various users for testing through use in real-life problem situations, and further feedback elicited.

The above outlines the various stages involved in the development of TRANZ. The intended users of this ES include personnel in highway agencies, construction companies, and others involved in the planning and design of work zones. Through use of this ES a uniform, rational practice in the area of traffic management strategies and choice of traffic control devices is sought.

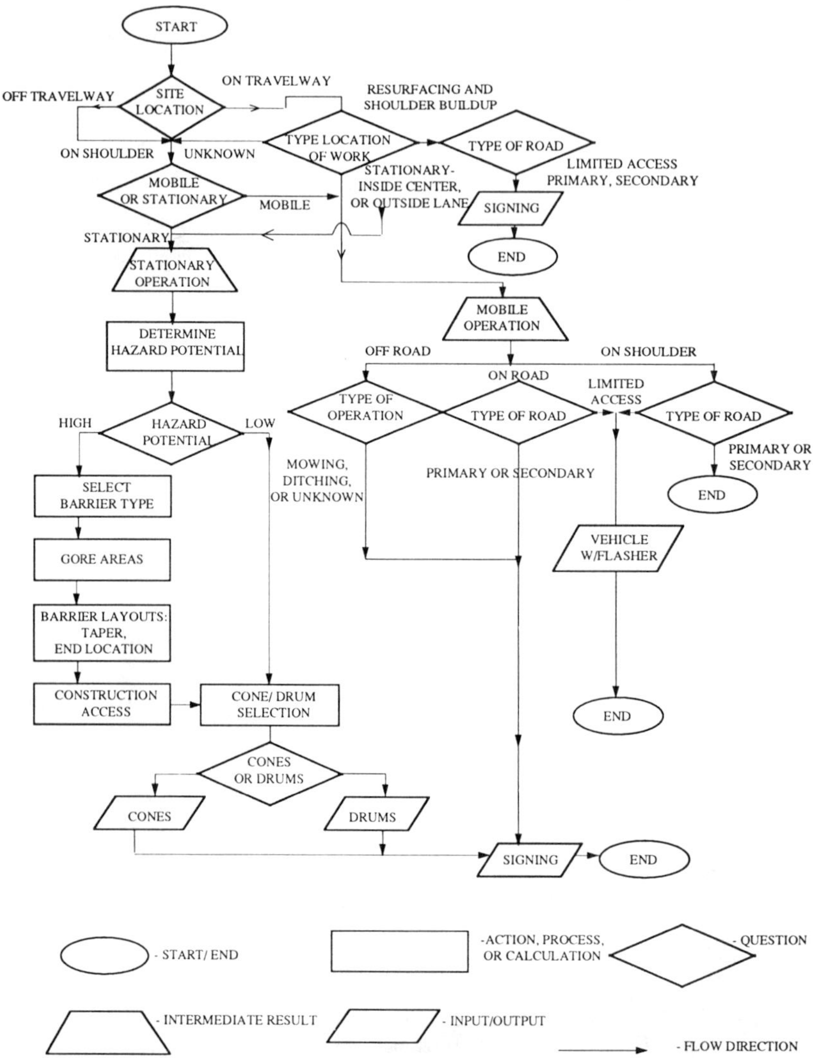

FIGURE 10. TRANZ decision framework. (From Faghri, A. and Demetsky, M. J., *J. Transp. Eng.*, 116(6), 759, 1990. With permission.)

6.5. EMERGENCY RESPONSE SYSTEM FOR DANGEROUS GOODS MOVEMENT VIA HIGHWAYS (Wilson et al., 1990)

Expert systems (ES) have been used in a large number of applications in transportation engineering. For an ES to offer a solution to a particular problem, the problem should be bounded and of reasonable size. It also must have relatively well-established rules. Application of ES methodology for emergency response procedures in accidents involving dangerous commodities is considered here. Numerous substances with a variety of hazards are transported over highways. Emergency response expertise and capability vary across different areas. The need exists for a system to provide a means by which the hazardous goods involved in an accident are identified, and uniform and correct safety precautions taken at the site.

The ES discussed here seeks to improve the quality of the dispatch information in an emergency situation involving dangerous goods (Wilson et al., 1990). The system is a user friendly program, designed for use by dispatchers who would receive information about an accident from a layman arriving at the scene. By making queries regarding the various aspects of the accident from the initial respondent, the dispatcher should be able to obtain the necessary basic information for input to the program. The ES was developed with the aid of an ES development shell called EXSYS. The following steps were used in creating the intelligent emergency response system:

1. Data collection and classification
2. Rule construction
3. System implemention

6.5.1. DATA COLLECTION AND CLASSIFICATION

During the data collection phase, truck movements were recorded at various survey points within the study area. Information regarding the origins and destinations of the trucks, the type of truck and the commodity it carried, payload of the truck, etc. was gathered and stored in a database. Using the information about origins and destinations from the database, a route map of the dangerous commodities being transported on arterial highways within the study area was prepared. The types of commodities were formed into groups as follows (Wilson et al., 1990):

- Chemical elements (excluding radioactive)
- Inorganic acids and oxygen compounds
- Inorganic bases and metallic oxides
- Metallic salts and peroxysalts
- Alcohols and derivations
- Explosives, fuses, and caps
- Chemical products in fertilizers
- Agriculture chemicals
- Paints and related paint products
- Other chemical specialties, industrial
- Gasoline
- Fuel oil
- Refined and manufactured gasses, fuel type
- Petroleum and coal products

The probability of a particular commodity group being present in the shipment is calculated by dividing the total number of shipments for that commodity group by the total number of movements for that particular location and vehicle type.

6.5.2. RULE CONSTRUCTION

ES make their decisions based on specific rules that are followed for specific information received. The rules within the ES discussed here are of three types: those relating to (1) the placard of the vehicle, (2) location and vehicle types, and (3) the type of shipping container and characteristics of the goods.

6.5.2.1. RULES RELATED TO PLACARD

Based upon the observations made to the dispatcher by the initial respondent regarding the placard on the vehicle, these rules were constructed such that for every combination of color and symbol on the placard, an appropriate dangerous-goods class can be selected by the ES. An example for a placard-related rule is listed below (Wilson et al., 1990):

IF			Placard color is white
		and	White symbol is skull and crossbones:
THEN	1.		Classification is Division 2.3
	2.	and	Classification is Division 6.1
	3.	and	Keep unnecessary people away
	4.	and	Keep upwind; isolate hazard area
	5.	and	Wear self-contained breathing apparatus and full protective clothing
	6.	and	Leak or spill
	7.	and	Eliminate all ignition sources

6.5.2.2. RULES RELATED TO LOCATION AND VEHICLE TYPE

Based on the location of the accident and type of vehicle involved, the dangerous commodities were separated into different groups and assigned a probability value. The probability value for a particular commodity being found at a particular location is calculated by dividing the original probability (assigned earlier from the placard rule) by the sum of the probabilities of all commodities in that group, for that combination of location and vehicle type. An example of a location-related rule is shown below (Wilson et al., 1990):

IF			Location of vehicle is near Woodstock
		and	Vehicle type is tractor-trailer
		and	Classification is Division 2.4:
THEN			
	1.		Chlorine—probability = 75/100

2. and Ammonia, anhydrous—probability = 13/100
3. and Ammonia, aqua—probability = 13/100
4. and Member of Division 2.4—probability = 3/100

6.5.2.3. Rules Related to Type of Container and Characteristics of Goods Within

A final set of rules are based on the type of container being used, and the odor, color, and physical state of the goods being transported. An example illustrates one such rule (Wilson et al., 1990):

IF		Commodity is contained in barrels:
THEN	1.	Arsenic trioxide—probability = 17/100
	2.	Ammonium chloride—probability = 17/100
	3.	Sodium hydrosulfite—probability = 17/100
	4.	Sodium phosphate (di)—probability = 17/100
	5.	Sodium phosphate (tri)—probability = 17/100
	6.	Ammonium nitrogen/phosphate for fertilizer—probability = 17/100

6.5.3. SYSTEM IMPLEMENTATION

An expert system (ES) shell called EXSYS was used for implementing the program. The program was designed to be extremely user friendly and simple to use. The user is presented with a number of screens and queried regarding the various characteristics. All acceptable replies, i.e., user choices, are numbered, allowing the user to input the appropriate response quickly. After all data are input, an analysis is done based on the rules. The program then displays the given conditions, a complete list of possible commodities, arranged in order from the highest to the lowest probability, and the appropriate list of emergency procedures.

After the trained emergency response personnel arrive on the scene of the accident, they may be able to provide further information about the various characteristics of the accident. In this event, the input conditions given to the system can be updated. The system would then reevaluate the new set of conditions and present a more focused set of commodities and emergency response procedures.

Thus, the ES methodology as implemented in this application provides a logical and reliable approach to the problem of handling dangerous goods in the event of a mishap during transportation. Possible advances by way of upgrades to the systems could include more information in the knowledge base, such as the environmental impact on the area, perceived danger to the available water and drainage system, agriculture and livestock in the vicinity of the accident, or any other area-specific information.

6.6. DEVELOPMENT OF A PROTOTYPE EXPERT SYSTEM FOR ROADSIDE SAFETY (Zhou and Layton 1991)

The collision of a vehicle with fixed objects adjacent to the roadway constitutes a sizable and critical part of the highway safety problem. To prevent or lessen the impact of such accidents, a traffic barrier to shield the fixed object is a possible solution. However, the barrier itself constitutes a fixed object, and hence the best location for the traffic barrier and the best type of barrier become vital issues in the context of roadside safety. In order to keep motorists on the travelway, various design features such as horizontal and vertical curvatures, lane and shoulder widths, signs and pavement markings, and various roadside safety features are employed, such as guardrails and breakaway supports. Once a vehicle leaves the roadway, the nature of the roadside environment is a significant factor in the probability of an accident occurring. A safe roadside environment for a vehicle that leaves the road can be provided by a clear zone, alongside the road, that is free of obstacles such as trees, utility poles, steep slopes, or lighting supports.

Some of the various options available to the designer for the treatment of roadside obstacles are (Zhou and Layton, 1991):

1. Remove the hazard or redesign it so that it can be safely traversed.
2. Relocate the hazard to a point where it is less likely to be struck.
3. Minimize the hazard by using an appropriate breakaway device.
4. Shield the hazard with a longitudinal traffic barrier and/or crash cushion.
5. Delineate the hazard, if none of these alternatives are feasible or practical.

Hazards on the roadside such as rock cliffs and permanent bodies of water cannot be removed or relocated. For such cases, traffic barriers such as guardrails are employed. The use of guardrails is warranted in situations where the consequences of hitting the guardrail are less than running off the roadway onto the hazard that the guardrail is shielding.

ES technology as an approach to problem solution is designed to provide the level of performance of a team of human experts and, through a computer, to assist people in varied locations to analyze specific problems using that expertise. A prototype knowledge-based ES for roadside safety is discussed here (Zhou and Layton, 1991). This ES (called ROADSIDE) evaluates whether a traffic barrier is necessary for a particular site.

The development of this ES can be followed in the following steps: (1) knowledge (data) acquisition phase, (2) rule-based representation, (3) system programming, (4) knowledge base structure, and (5) inference engine.

6.6.1. DATA ACQUISITION

In this phase, problem-solving expertise from knowledge sources such as research reports and design guides is gathered and stored in a database of the ES. These data can be supplemented by results from ongoing research.

6.6.2. RULE-BASED REPRESENTATION

The knowledge in the database is organized and represented by using rules. In the ES ROADSIDE, the knowledge representation is achieved through a backward-chaining format. In a backward-chaining format, the system starts with the conclusion and proceeds back to the initial condition. Examples of rules in this ES are (Zhou and Layton, 1991):

- Rule 1: Guardrail may be warranted:
  If a location is nontraversible
  Or a fixed object is present.
- Rule 2: The location is nontraversible:
  If there is an obstacle within the clear zone
  And the obstacle cannot be removed or relocated.

Knowledge representation in this ES is done through rules that follow a backward-chaining format. In a backward-chaining implementation of rules, the reasoning process starts from a conclusion and works backwards to the conditions specified at the input, i.e., in a reverse fashion from typical reasoning. Figure 11 shows the simplified reasoning process in the ES ROADSIDE.

6.6.3. SYSTEM PROGRAMMING

This ES was developed using the artificial intelligence (AI) programming language Prolog. The use of Prolog provides greater flexibility and adaptability in developing the system. Prolog also has a built-in compiler, so that a program can be executed directly as a stand-alone program with no need for a system development environment. The cost of the Turbo Prolog program is inexpensive compared to an ES shell, which generally costs a few hundred dollars. The disadvantage of using programming languages for the complete development of an ES, however, is that one has to learn thoroughly the programming language first, and then start writing the process of coding. Debugging the program is often more difficult and time consuming, since a considerably larger number of logical and syntactic errors could be encountered compared to development through an ES shell.

6.6.4. KNOWLEDGE BASE STRUCTURE

The knowledge base is the collection of domain knowledge, as imparted by the expert, or knowledge engineer to the system. The knowledge base of

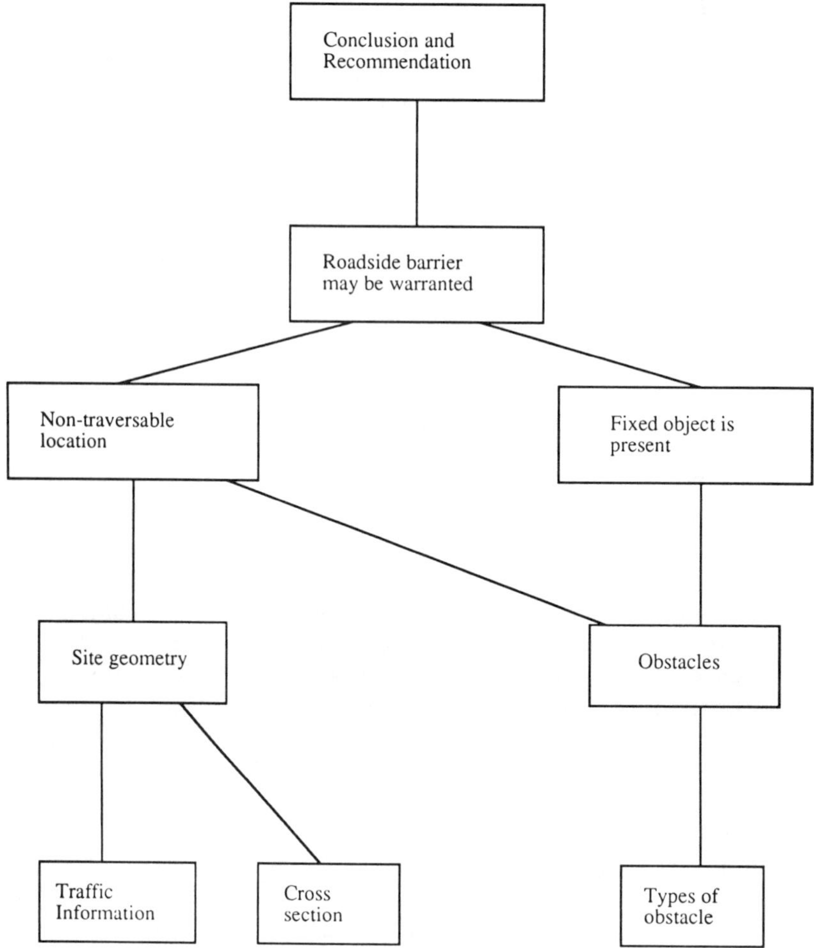

FIGURE 11. Simplified reasoning process. (Zhou, H. and Layton, R. D., *J. Transp. Eng.*, 117(4), 437, 1991. With permission.)

ROADSIDE is designed as an individual unit that can be edited using a text editor. This independent unit can be updated or revised with the addition of a further set of facts. The knowledge base consists of a rule base and a working memory. The rule base holds the rules and facts acquired by the ES, and the working memory stores the facts or data obtained from the user during program use.

6.6.5. INFERENCE ENGINE

The basic formal reasoning is called inference. Inference involves matching and unification by means of rules. The inference engine takes the facts that it knows to be true and relevant from the ''rule base'', and the working memory of the computer uses these facts to test the rules in the knowledge base through the process of unification.

A prototype ES for roadside safety analysis has been analyzed. This system was developed in particular to evaluate whether a traffic barrier is necessary for a particular site being analyzed. Through development of the prototype, the applicability of ES in the field of highway safety analysis has been demonstrated.

6.7. ADVISORY SYSTEM FOR DESIGN OF HIGHWAY SAFETY STRUCTURES (Roschke, 1991)

Among the efforts being made in recent years to improve highway safety, major emphasis has been placed on the nature of the roadside environment. Fixed obstacles adjacent to the roadway contribute traffic hazards; this often can be remedied only through the removal of these hazards or by shielding them by the use of traffic barriers.

A knowledge-based expert system (KBES) for the selection, location, and design of traffic barriers such as guardrails, crash cushions, and similiar roadside barriers is discussed here (Roschke, 1991). The ES (called SAFEROAD) aids in identification of sites that warrant barrier protection. It also aids in the choice of the type of barrier, based on strength, safety, maintenance requirements, and cost. The architecture of SAFEROAD is illustrated in Figure 12 (Roschke, 1991).

The following units constitute the major components of this ES: (1) knowledge base, (2) inference engine and rules, and (3) user interface.

6.7.1. KNOWLEDGE BASE

The structure of the knowledge base is in Figure 13 (Roschke, 1991). The knowledge base is structured in a modular fashion. The highway safety structures are subdivided into four categories, with a knowledge base containing information about each category. Thus, there are separate databases on crash cushions, roadside barrier, median barrier, and bridge rails. Each category is further divided into two subset knowledge bases, one for retrofit constructions and another for new constructions. These eight knowledge modules are directed by a controller knowledge base.

6.7.2. INFERENCE ENGINE AND RULES

SAFEROAD was developed using the ES shell or development tool ''NEXPERT Object''. This shell provides inference mechanisms, rule formats, and an editor environment for the ES program to be developed. The rules can be processed in either a forward-chaining or backward-chaining fashion, using the same format. SAFEROAD has an explanation facility by which the reasoning process, and the graphical and written directions for

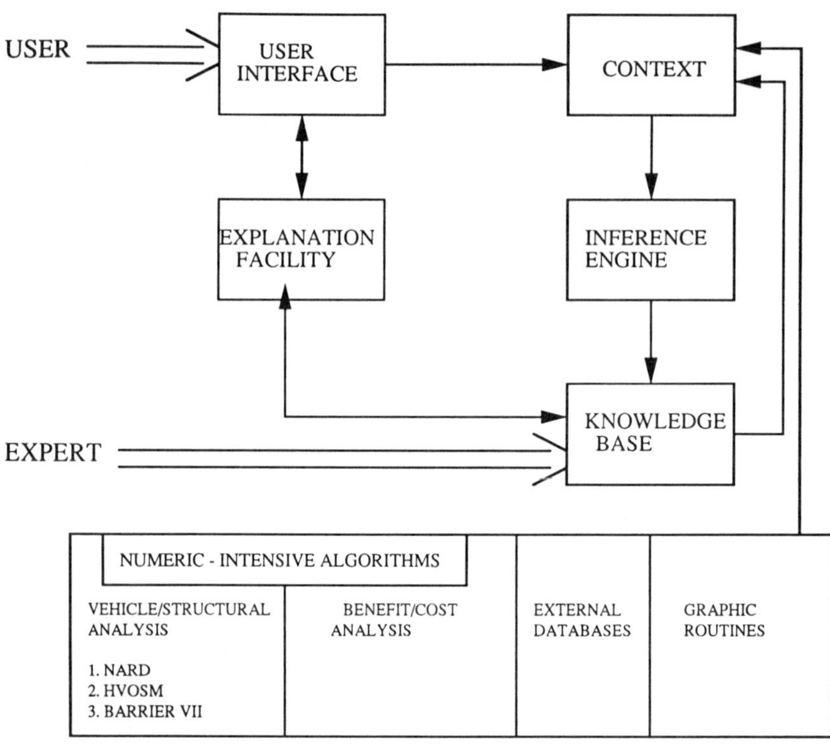

FIGURE 12. System architecture. (From Roschke, P. N., *J. Transp. Eng.*, 117(4), 421, 1991. With permission.)

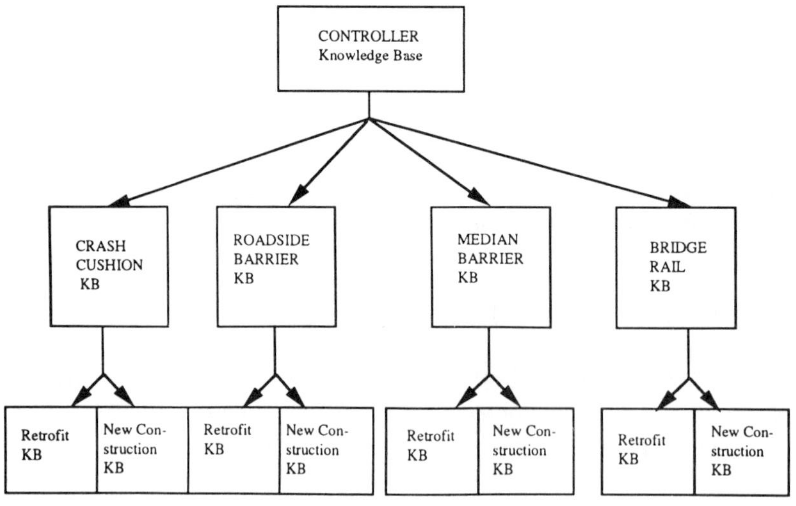

FIGURE 13. Modular structure of knowledge base. (From Roschke, P. N., *J. Transp. Eng.*, 117(4), 421, 1991. With permission.)

traffic barrier installation and design can be accessed by the user. Thus, once a recommendation is established, such as ''design crash cushion'', the user can request an explanation session, and by backtracking of the recorded rule-firing sequence, a list of fired rules in a chain of reasoning are made available. The benefit vs. cost analysis in SAFEROAD is done through an external routine feature provided by the development shell. A ratio between the benefits obtained and the costs of an improvement is used to determine if it is cost effective (Roschke, 1991):

$$BC_{2-1} = \frac{SC_1 - SC_2}{DC_2 - DC_1} \tag{2}$$

where:

BC_{2-1}	benefit-cost ratio of alternative 1 compared with alternative 2
SC_1	annualized societal cost of alternative 1
DC_1	annualized direct cost of alternative 1
SC_2	annualized societal cost of alternative 2
DC_2	annualized direct cost of alternative 2

Alternative 2 is usually considered to be an improvement over alternative 1. When the benefit/cost ratio for a safety improvement is below 1.0, the benefit usually is not implemented. In SAFEROAD, input information such as the accident costs for the benefit-cost procedure is read from a database. Output, i.e. the benefit/cost ratio of one alternative compared with other alternatives, is supplied to the inference engine and used in the evaluation phase.

6.7.3. USER INTERFACE

SAFEROAD is designed to have a user-friendly graphical interface through the computer screen. A main menu collects all system functions into a pull-down menu format for selection of problem tasks such as file manipulation and explanation functions. The user is presented with various choices and is required to type in the choice number in order to make a selection.

6.7.4. STRUCTURE OF THE KNOWLEDGE BASE

The knowledge-representation method in the expert system SAFEROAD can be classified as belonging to three categories: (1) static knowledge, which is defined as that knowledge representing physical structure, along with its components and their topology, (2) dynamic knowledge, which is the knowledge of design constraints and heuristic laws that are to be satisfied in a given problem, and (3) graphical knowledge, which is largely visual knowledge.

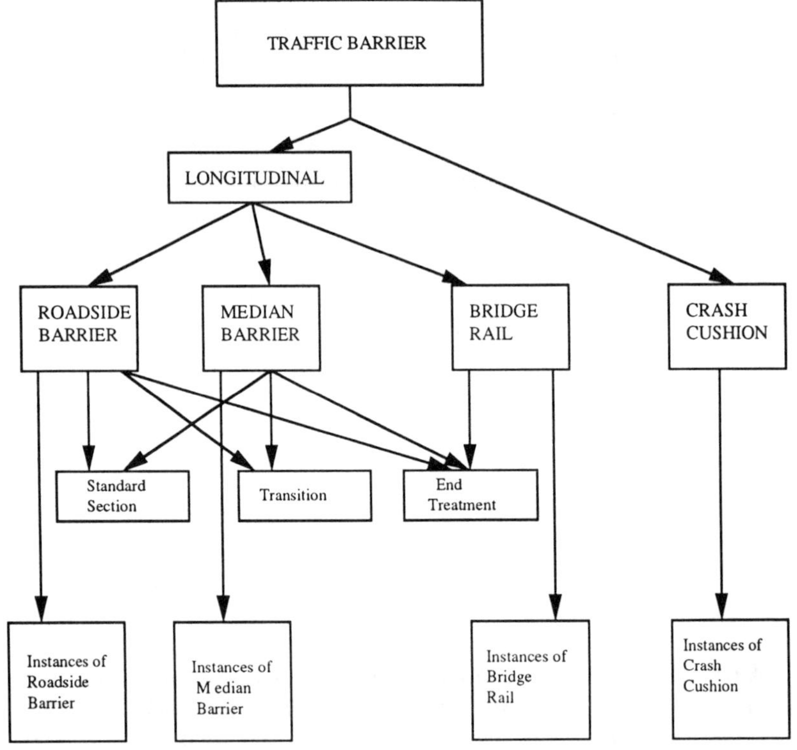

FIGURE 14. Taxonomy of highway traffic barriers. (From Roschke, P. N., *J. Transp. Eng.*, 117(4), 425, 1991. With permission.)

6.7.4.1. Static Knowledge

Static knowledge in SAFEROAD is organized as follows. First, the traffic barriers are classified. Four main groups, such as roadside barriers, median barriers, bridge rails, and crash cushions, are listed at the primary level. Next, these barriers are subdivided into specific instances of devices installed along roadways. This structure is shown in Figure 14 (Roschke, 1991). At the first level, four major kinds of traffic barriers (e.g., roadside barrier, median barrier, bridge rails, and crash cushions) are grouped together. Specific instances for devices are considered at the next level. The taxonomy of crash cushions is depicted in Figure 15 (Roschke, 1991).

6.7.4.2. Dynamic Knowledge

Dynamic knowledge is represented by production rules. The rules in SAFEROAD have a left-hand side (LHS) and a right-hand side (RHS). The LHS contains the condition that must be met for the rule to be considered true. The RHS contains the rule conclusion and other actions that are to take

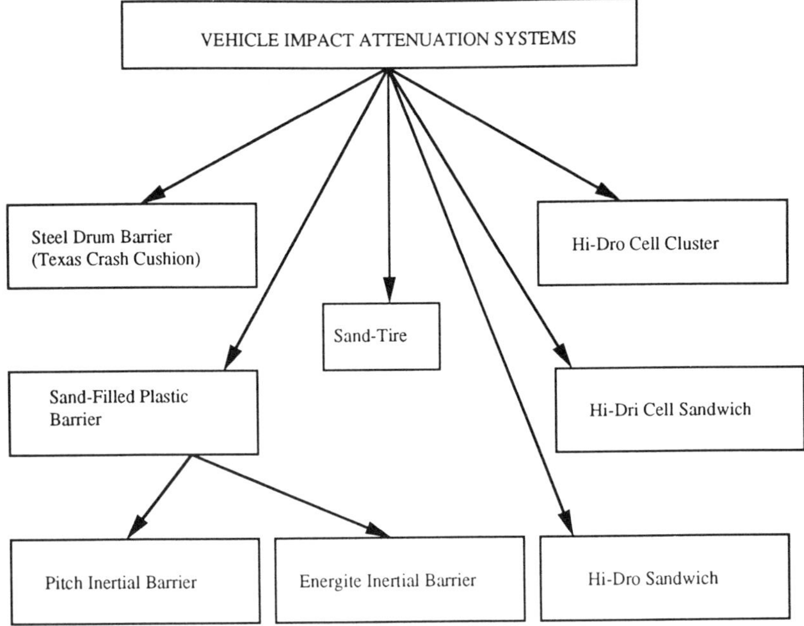

FIGURE 15. Taxonomy of vehicle-impact-attenuation systems. (From Roschke, P. N., *J. Transp. Eng.*, 117(4), 425, 1991. With permission.)

effect if the rule is found to be true. The inference engine performs pattern matching when trying to make a list of objects satisfying a given set of conditions.

6.7.4.3. Graphical Knowledge

The graphical knowledge in SAFEROAD is created through a document-scanning process. Figure 16 (Roschke, 1991) illustrates the steps involved in the acquisition of this visual knowledge. First, documents in the form of photographs to line drawings are read into memory through a scanning machine. After a document is scanned, it is given a file name with the extension ".MSP" for a black and white picture or ".PIC" for color. The inference process can then display the image file on the screen when called for in the program.

6.7.5. WEIGHT-ASSIGNING MECHANISM

The large number of traffic barriers available, and the complex decision factors involved with each possible design, can present a problem to the engineer. A mechanism of assigning a weight of importance to each decision factor in SAFEROAD was developed. This mechanism evaluates the positive or negative effect of each factor, thereby evaluating and ranking the various

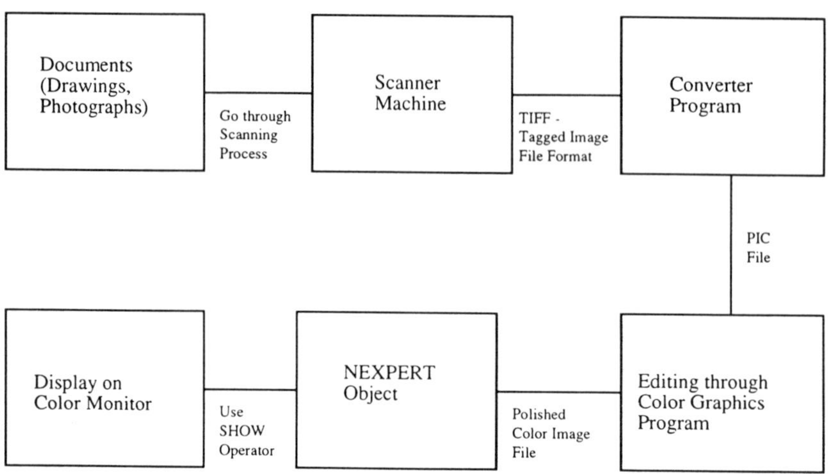

FIGURE 16. Acquisition of visual knowledge. (From Roschke, P. N., *J. Transp. Eng.*, 117(4), 428, 1991. With permission.)

designs according to desirability of choice. A given traffic barrier design is assigned a "weight of importance" value and a "weight of preference" value. The combination of these two values leads to a "total preference value", which represents how the given design ranks on a scale of -1 to $+1$. Thus, a complete evaluation of decision factors for the selection of an optimal design reaches the most suitable design by a comparison of these values for each option. The higher the total preference value, the more desirable its respective traffic barrier design for a particular location.

The KBES, SAFEROAD, was developed using the ES development shell "NEXPERT object". SAFEROAD was designed with the aim of helping engineers in selecting, locating, and designing traffic barriers in both new and retrofits of existing projects. Through incorporation of visual knowledge preserved through scanned images that can be viewed on the computer screen, the system offers very good educational and training benefits to highway engineers. The structure of the knowledge base provides for updating and revising the facts contained therein, protecting against system obsolescence.

Chapter 7

EXPERT SYSTEMS IN ENVIRONMENTAL AND WATER RESOURCES ENGINEERING

7.1. INTRODUCTION

Several applications for expert systems (ES) have been identified in the fields of environmental engineering and water resources engineering. The problems of hazardous waste management, i.e., the disposal of industrial or other wastes so as to have minimal impact on the environment, are of particular interest. Examples of specific problems in environmental engineering for which the ES approach has been considered include: ES to guide reviewers of hazardous waste facility permit applications, waste facility siting, monitoring monthly levels of waste generated by sites in order to enforce regulatory limits, analyze unknown containers of hazardous waste, aid in identification and suggest storage and disposal methods for hazardous wastes, dredge material disposal, etc.

Problems in the area of water resources engineering that have received consideration through the ES approach include: ES to interpret groundwater sample data, to assist filter plant operators in problem diagnosis, remedial suggestions, and to improve the operation of processes, in water treatment plants to schedule pumping operations based on projected water demand and energy costs, inform plant operators about the quality of the water and suggest required maintenance procedures, in reservoir siting and reservoir management to maximize water supply and hydropower output, etc.

Research on these and several other systems reflects the strong interest in, and applicability of, ES technology in the field of environmental and water resources engineering.

7.2. EXPERT SYSTEM FOR HAZARDOUS SITE EVALUATION (Wilson et al., 1987)

Improper disposal of hazardous and toxic substances is the major cause of pollution. Assessing the problem of waste disposal methods and the contamination potential of a waste disposal facility is a complex one, requiring knowledge and information from toxicology, chemistry, and environmental engineering, among other disciplines. In the case of large or important waste disposal sites, an overall assessment is usually conducted by a team of experts who assess the impacts of the diverse aspects of the pollution problem. However, only one geologist or environmental/civil engineer is involved in most instances of site evaluation. In such cases, it is desirable to have an ES by which the collective opinion of the experts is accessible to the engineer to assist him in decision making. An ES (called GEOTOX) designed for the evaluation of hazardous sites is discussed here (Wilson et al., 1987).

GEOTOX is a knowledge-based expert system (KBES) designed for evaluating a hazardous waste site. It is specifically intended to be of assistance in preliminary investigations, and can be used for site comparison, prioritization, and ranking. Apart from the evaluation of existing sites, GEOTOX can be used to assess potential sites and assist in the evaluation of the site selection process for new facilities. The various processes of GEOTOX can be summarized as:

1. Interpretation: the assessment of existing hazardous waste sites
2. Classification: ranking of existing sites, screening of potential sites
3. Diagnosis: contamination problems at hazardous waste sites, selection of remedial alternatives

Thus, consultation with GEOTOX assists the user in task interpretation by organizing and classifying the data. The inference capabilities of GEOTOX allow the system to be used in diagnostic tasks as well. The system can identify the potential problems at a site, the contamination possibility and type of contamination, and the seriousness of the situation. By applying GEOTOX and varying the input data, the evaluator can study different alternatives for remedial action, depending on the site conditions.

Assessment of hazardous waste sites requires several elements of probabilistic reasoning: the subjective judgement of expert professionals is also an important component in the evaluation process. GEOTOX was designed around the typical framework for an ES, containing (1) a knowledge base (holding the rules and application knowledge), (2) the databases, each related to a specific domain and containing the classification and types of contaminants or soils etc., (3) the inference mechanism for knowledge processing and modification, (4) algorithmic structures and analysis programs, and (5) a user interface, which communicates with the user.

The stages of development in GEOTOX can be classified as below:

1. The problem was first characterized, long-range goals defined, and data acquired.
2. The key concepts and relationships of stage 1 were formalized, and organized into a hierarchical structure, to form the conceptual basis for a prototype system.
3. The data structures, inference rules, and control strategies defined in the previous stages were implemented in a computer program.
4. The prototype program was evaluated, and revisions carried out if necessary.

7.2.1. CONCEPTUAL FRAMEWORK OF GEOTOX

The various components of GEOTOX can be studied as a framework consisting of the following main components:

1. The user, at the highest level, who operates and accesses the information in the system
2. The knowledge base, which stores the rules and the application knowledge
3. The databases, related to the domain
4. The inference mechanism, for knowledge processing and modification
5. The database, which stores the facts known by the user (known facts) and the facts deduced from the inference procedure
6. The algorithmic structures and analysis programs, to analyze the contaminant transport, groundwater flow, etc.
7. A graphics system plus a computer-aided design (CAD) package, through which containment alternatives (such as covers, liners, drainage systems, pumping wells, etc.) can be designed
8. The user interface, between the system and the user, through which operational processes and explanations are communicated to the user

7.2.2. ARCHITECTURE OF THE KNOWLEDGE BASE

The knowledge representation scheme in GEOTOX is based on an associative network of the relationships between the various parameters of the site. The associative network represents the generic problem principles. This network forms the basis of the knowledge representation scheme and defines all the associations between data and site parameters. Every site characteristic is represented by a "node" in the network. Production rules are attached to the conjunctive nodes and leaf nodes. These rules are used to determine the effects of particular input parameters on the overall characteristics of the site. Thus, the scheme consists of an associative network, with rules attached to the leaf nodes and the conjunction nodes. In concept, the associative network can be thought of as being able to define "what is going to affect what", and the production rules say "by how much". The expert-provided descriptors of situations are stored in the knowledge base in the form of "frames" attached to the higher nodes of the network. Thus, the associative network represents the generic problem principles, the rules represent the heuristics and expert reasoning, and the frames represent the expert-defined conditions.

7.2.3. EXPERT RULES

For each characteristic of the site, there is an associated set of rules that defines its contribution to the overall site hazard. In GEOTOX, two values, a hazard value and a confidence level, are used together in order to judge the effect of a given characteristic of the problem site on the overall hazard. The hazard value of the rule (h) is defined by the expert, and represents the indices for the severity of the situation at the site. The confidence level indicates the expert's confidence in the application and strength of the rule. The confidence level is similar to a weight related to the hazard index. Assigning the hazard-confidence values to a node, in effect, represents the use of expert rules of thumb to decide on the effects of the given conditions.

When evaluating a site, GEOTOX tries to resolve the ultimate question of ''How good, or how bad is the site''. The approach taken is to assign a hazard index, H, on a scale from 0 to 10 such that a higher H indicates a higher potential hazard from the site. An example to illustrate the concepts and methodology used in GEOTOX is given below (Wilson et al., 1987). The problem considered here involves a toxic chemical released into a stream. One must determine the contamination level of a water sample taken from some point downstream. If the concentration of the chemical at that point is known to be greater than a safe limit established from regulatory guidelines, then it is certain that the water is hazardous. However, if such data are not readily available, the water contamination level could be estimated by an expert on a scale of 0 to 10, depending on the distance from the source and the chemical concentration. This information can be expressed in the form of a rule and put into the knowledge base.

IF the distance is between 10 and 50 ft
AND the concentration is greater than 5 ppm
THEN the stream pollution hazard index (H) is 8

If the concentration level is unknown, then it is necessary to apply another expert rule, based on distance only, and include it in the knowledge base:

IF the distance is between 10 and 20 ft
THEN the overall stream pollution hazard index (H) is 8

At this stage, the rule could be modified in order to incorporate a measure of the expert's belief in the applicability of that rule. The confidence level of a rule is defined based on the probability of the overall hazard being H, given the hazard value for the rule, h. Thus, the rule would now become,

IF the distance is between 20 and 50 ft
THEN the overall hazard is 8 with a confidence level of 0.30

Thus, GEOTOX uses the hazard value of a rule (h) and the confidence level for that rule (c) and propagates them to all the associated nodes following the links in the associative network. The system asks for data corresponding to the leaf nodes of the network and then finds the applicable rule for the given data. The rule fires, and a hazard value-confidence level (h-c) pair is assigned to the corresponding leaf node of the associative network.

7.2.4. USER INTERFACE

The user interface in GEOTOX provides facilities by which the user can examine the contents of the knowledge base as well as the current state of consultation, and can ask "how" and "why" questions about the results (by tracing up and down the associative network, and by displaying the applicable rules, if any). The user can also volunteer information or change his response to previous questions.

Thus, the knowledge base and the user interface of GEOTOX make it a versatile and powerful tool for decision making in the identification and remediation of hazardous waste sites. Use of the ES approach to this problem provides the benefits of rapid inspection, better knowledge of the situation, improved assessment of the problem, and remedial decision making.

7.3. EXPERT SYSTEM FOR EVALUATING HAZARDOUS WASTE GENERATORS (Knowles et al., 1989)

The problem of monitoring and regulating the units or organizations responsible for releasing hazardous waste materials into the environment is one of growing size and complexity. The process of evaluating information on these waste generators and of intimating them in this regard can be expedited through the use of expert system (ES) methodology. An ES for the evaluation of hazardous waste generators is considered here, which also expresses the final technical result in the form of an action letter to the respective generators (Knowles et al., 1989).

This ES uses the Lotus 1–2–3 spreadsheet as a programming environment for its implementation. Lotus 1–2–3 is a widely used spreadsheet package for a variety of problems, ranging from normal calculations to doing graphics and database management. It is highly successful due to its ability to perform numerical processing. The ES model developed here is very similar to a system developed using a development shell. The only difference between this model and an ES is that an algorithmic control system is used instead of the inference-based control system.

The architecture of this ES model is shown in Figure 1 (Knowles et al., 1989). As seen from Figure 1, the model has six major components, each representing a separate level. The expert supplies the knowledge base through the knowledge acquisition level. The knowledge base can be updated with further information as required, and then searched based on the specifications made by the user in the rule base. The knowledge base and the rule base are evaluated by the context component, and the results used by the control system to process the correct output responses in the form of an action letter.

7.3.1. KNOWLEDGE BASE ORGANIZATION

The knowledge base consists of information coding and look-up tables. The hazardous waste types are placed into 26 waste categories, and each is

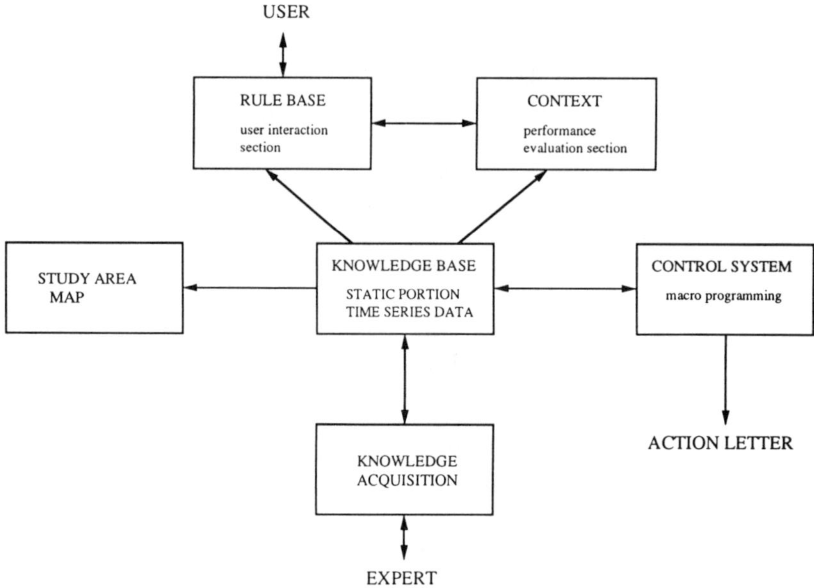

FIGURE 1. Architecture of action letter model. (From Knowles, L. et al., *J. Comput. Civ. Eng.*, 3(2), 111, 1989. With permission.)

assigned a letter of the alphabet. An American Standard Code for Information Interchange (ASCII code) is given to each of the 26 letters. By providing these codes in a look-up table, the table can be referenced from another portion of the spreadsheet during the information search and extraction processes. The look-up table is created through Lotus such that the first column contains the ASCII code, the second is the waste category, and two more columns are designated for "hazardous waste type" and "regulation limit". The knowledge base is organized such that all information is represented in an object-attribute-value format. Objects are represented by the names of the organizations or sites generating the waste, and tagging the attributes for waste type and amount generated. This format allows quick scanning of the database for values that meet prespecified criteria. Once a value is found, any attribute connected to it can also be extracted, and listed for user observation and further evaluation. The fixed-data section of the knowledge base consists of an inventory of hazardous wastes. Data regarding the waste generator is organized and represented such that the first column contains the name of the site/person, and subsequent columns contain attributes of address, waste type, waste amount, type of storage employed at the site, and disposal method. The time series data section of the knowledge base contains the monthly waste generation quantities. Attributes in this section are identified by month names (Jan., Feb., etc.) and contain the values of the waste amounts for that month. Graphs can be obtained, and trends observed, for waste generation and disposal. Thus, this form of data organization is facilitated by the very nature of the Lotus spreadsheet.

7.3.2. PERFORMANCE EVALUATION SECTION

In this section, the information in the knowledge base is evaluated using the selected criteria. The performance evaluation section identifies those sites that exceed a maximum quantity of hazardous waste generated per month and/or use specific storage/disposal methods. The evaluation section is controlled by the rule base. The performance evaluation section was created by writing a formula in each cell to perform a database inquiry within that same row. If, for example, quantity violations are evaluated in the first column, the formula searches for the regulation limit identified with the waste code in the fixed-site data. Other columns in the evaluation section can also be set up as cell formulas for scanning the time series section of the knowledge base according to specifications in the rule base.

7.3.3. DEVELOPMENT OF THE RULE BASE

A set of rules are defined, and the user can search data within the database based on these rules. Additional rules can easily be inserted into the system to meet future needs. Here, the rule base has been subdivided into three categories: violation rule, and rules for disposal and storage methods. The violation rule identifies sites exceeding a regulatory standard, and hence all sites that exceed a maximum monthly quantity will be extracted from the database and listed in an extraction table. Through the "disposal methods" category, the database can be scanned for generators using any of 15 possible disposal practices, and the corresponding sites extracted. Similarly, the "storage method" category allows for search involving up to ten different storage practices. Once all the options have been selected, the control system applies the rules to the database search, extraction, and listing processes.

7.3.4. CONTROL SYSTEM

The control system of the spreadsheet is based on a programming macrosequence. It controls the flow process of the analysis in a forward-chaining operation. After the search criteria have been specified, the user initiates the search process, the extraction process, and the letter-processing operation. If a user specifies a search for quantity violators, the information in the knowledge base is analyzed and compared to the maximum values in the look-up table. If the values are exceeded in the database, specific information is extracted into a table. From the table, an appropriate message is also selected, to be presented in the form of a letter to that waste generator. Thus, the flow logic process of the control system involves rule category, rule base search, data extraction, message selection, and, finally, action letter development.

7.3.5. PROGRAMMING AND DEVELOPMENT

The entire system was developed using Lotus 1–2–3 as the single programming tool, which greatly simplified the development and programming process. Among other features offered by the spreadsheet, characters or words in different spreadsheet cells can be added together to create phrases and sentences, through string arithmetic. As a consequence, no ES programming "shells" or aids were required.

The ES described here serves to expedite the process of both evaluating the information on sites responsible for generating hazardous waste and reporting the results to them. Thus, the system makes technical analysis more efficient and reduces the time required to produce the associated paperwork.

7.4. AN EXPERT SYSTEM FOR COMBATTING OIL SPILL POLLUTION (Orhun and Demirors, 1991)

In incidents involving oil spills at sea, it is crucial to provide a timely and adequate response in order to minimize damage to the environment. If emergency action is not taken during the first few hours after the spill, more oil is lost to the environment, and this delay can cause the oil to spread over a greater area with the consequence that longer lengths of coastline are affected, resulting in more expensive cleanup operations. The emergency action procedures taken to minimize the damage done to the environment are termed oil-spill response. The cost of cleanup operations is usually high, and is an uncertain parameter. Depending on the availability of technology, its suitability for use, the skills and training of the people responsible for combatting the spill, etc., a decison is made as to the technique to be implemented for effective handling of the spill.

The basic techniques for dealing with oil-spill responses are classified as follows:

1. Mechanical spill response techniques, such as booms and skimmers
2. Dispersants (chemicals to disperse the oil)
3. Burning
4. Bioremediation (the local use of microbes to biodegrade and oxidize hydrocarbon molecules)

Analytical techniques and approaches to combatting the problem of oil spills offer limited support in the handling of real-life decision-making situations with incomplete or incorrect data. The analytical techniques or procedural and reference guides also do not offer any kind of interactive support to the user. Through use of ES technology, the domain-specific knowledge of experts in the field can be consolidated and used to solve problems. The expert is capable of using incomplete or uncertain information and, through a process of reasoning and by applying heuristics, can narrow the solution range. The characteristics of the oil-spill problem make it very suitable for approach through the ES method. The ES discussed here shows how technology is used for decision support in combatting oil spills. First, a decision needs to be made based on availability, i.e., how far away from the problem site the cleanup equipment is stored, the quantity and types of equipment, etc. At the next level, using this information as one of the criteria, the spill response action for the site, such as what equipment should be dispatched to the scene and in what quantities, etc., is determined. Finally, the operational level examines in detail the actions that must be taken at the site, such as where to deploy the boom, etc. Computer programs are developed to support decision making at each level. For the first consideration, inputs would include data such as equipment performance and cost, weather, spill trajectory, legal situation, etc., in order to generate outputs to aid in the decision making. The problem of locating emergency service facilities such that all possible demand sources can be covered, while keeping the number of facilities to a minimum, is also considered in terms of "maximal covering" models. The maximal covering models developed for fire station/ambulance location problems could be extended to oil-spill resource siting. For the next level, the optimal dispatching of cleanup equipment is modeled. Inputs to this model would include data such as the outflow of the oil spill, cost of transportation, etc., with the goal of minimizing a weighted combination of response and damage costs. The decision-making process is carried out in the presence of several conflicting objectives. Some of the attributes of the problem (e.g., cost of damage, cleanup, etc.) can be measured quantitatively, while others (e.g., aesthetics) are highly subjective in nature. A decision is sought by maximizing the vector of objectives, subject to the constraints on the decision variables. An optimal decision is then defined as one that maximizes the utility function of the decision maker.

The prototype ES discussed here was developed using the ES shell, Personal Consultant Plus (PC+). The components of this ES include a knowledge base, a rule base, and a user interface. The knowledge base is the repository of the domain-specific knowledge provided by the expert, and is stored in a large database that can be referenced by the rules and the inference mechanism. The general structure of a rule is:

IF <condition> THEN <action>

The conditions and actions contain parameters, values, or functions. Rules involving parameters define how the value of the parameter would be obtained. An example of a rule for handling chemical dispersants is shown below (Orhun et al., 1991):

IF chemical dispersion is a solution
and personnel and equipment for chemical dispersion are available

	and weather conditions for chemical dispersion are suitable
	and dispersion is acceptable environmentally
THEN	disperse CF = 90

The various states and properties of facts are stored in the knowledge base, using parameters. Based on the value that the parameter is given, e.g., ask-all, single value, multivalue, and yes/no, there are four kinds of parameters. If the value of the parameter can be derived by querying the user, the parameter is phrased in the form of a question, as shown below.

THE SPILL MATERIAL	
TRANSLATION	(kind of spill material)
PROMPT	What is the type of spilled material?
RESPONSE	SINGLEVALUE
EXPECT	volatile product, light refined product, heavy refined product
USED BY	RULE009 RULE010 RULE013 RULE014 RULE035
UPDATED BY	RULE041 RULE042 RULE045

Both rules and parameters can have certainty factors associated with their actions. Certainty factors can be defined for each rule or parameter, and take a value between −100 and +100.

In the PC+ ES shell, information needed to solve a specific problem is organized using "frames", which group or categorize a collection of knowledge that is characterized by similar attributes or related parameters and used together as a single unit. Each frame of a problem groups together the various parameters and rules required to solve that problem. Thus, frames describe a specific problem area within the domain of an ES. The frames are organized in a hierarchical fashion such that the resolution of a problem into subproblems can be depicted. This structure is called a frame tree. Figure 2 (Orhun et al., 1991) shows a part of the frame tree of the ES considered.

The user interface of the ES provides for an active interaction between the user and the system, through the keyboard and the display unit of the computer. When the ES is invoked for consultation, the domain is displayed, and the user can commence the session. Based on the inputs received from the user, the predefined goals of the frames are satisfied by conducting inferences, starting from the root and applying the rules and querying the values of parameters as required, to reach the conclusion, which could be the suitable techniques for combatting the oil spill, the equipment required, actions to be taken, etc. A feature of this ES is that the user can at any point question as to "what", "how", or "why" the conclusion was reached and receive a review and explanations to that effect. By incorporation of this query or review, the oil pollution combat system can be used in training, as an educational tool as well as for decision support. Another use for the system could

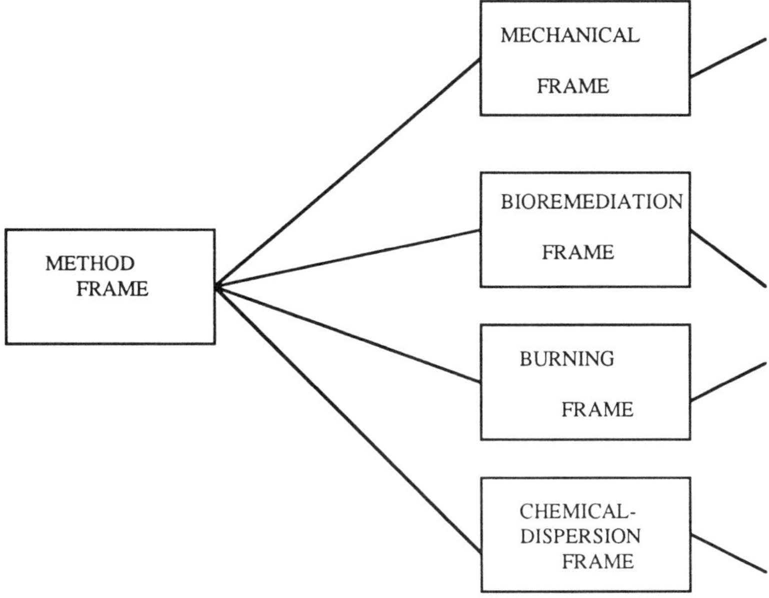

FIGURE 2. A subset of frame-tree used. (From Orhun, E. and Demirors, O., *14th Annu. Energy Sources Technol. Conf. Exhib.*, ASME, New York, 1991. With permission.)

be depiction of various spill scenarios in critical areas, so that a level of preparedness can be achieved for solving oil-spill issues.

7.5. AN EXPERT SYSTEM FOR AIDING IN PESTICIDE REGULATORY DECISIONS (Crowe and Mutch, 1990)

Pesticides are poisons introduced into the environment, with the intention of killing target species of plants or organisms. Pesticides serve to enhance crop production and quality. The risks to the environment associated with the use of these pesticides are considerable, and hence the quantity and method of use of these pesticides are governed by regulation. However, even if the recommended application procedures are followed, the hazard to the environment cannot be ruled out. For example, strong evidence links the pesticide Aldicarb with the contamination of groundwater. A complex set of chemical, biological, and physical processes govern the transport and transformation of pesticides in the unsaturated soil zone, requiring access to and application of specialized knowledge. There is a need to develop a model for simulating the major processes controlling the eventual destination, or fate, of pesticides that can be easily used by regulatory personnel assigned the task of assessing the effects of a pesticide on the quality of the groundwater. A complex problem requiring the application of a numerical model can be approached as follows:

In the conventional method, the person with the problem visits an expert, such as a numerical modeler, who has knowledge about the numerical model of the problem as well as the field of application. The customer and modeler discuss the available data and the purpose of the study. The expert prepares the input data set in the appropriate numerical code, runs the model, and uses the results of the simulations as a basis for a conclusion.

In ES methodology, the same steps are involved, except that the expert(s) need not be approached. Instead, they provide the body of their knowledge in a repository that is part of the ES. The ES would converse interactively with the customer in order to obtain input from him, evaluate the data, and convey the results, again through an interactive session with the user, this time with the system providing explanations and illustrating the reasoning process. A properly constructed ES can duplicate the knowledge of human experts, with the advantages of reduction in time taken for obtaining the results, reduction in costs, notably the fees of the expert(s) involved in the consultation. An additional advantage is that an ES can be used for training purposes, i.e., as an educational tool for study and analysis of problems in a domain.

ES typically possess large database that hold information as well as data. Information is of two types, facts and knowledge. Facts include data obtained from literature surveys of textbooks, manuals, procedural guides, etc. Knowledge includes a set of facts, thumb-rule convictions, and procedures or rules for solving a problem. Heuristic knowledge is derived from experience gained through solving problems in the past.

ES are also commonly composed of a combination of rules and frames. ES based on production rules are known as rule-based systems. The rules in ES are of the form,

IF <condition> THEN <action>

i.e., if the condition or premise stated in the IF part of the rule is satisfied by the data collected, the THEN part of the rule (action) is executed. An example of the IF-THEN structure of a production rule is given below (Crowe and Mutch, 1990):

IF Aldicarb is detected in the groundwater
THEN the water is contaminated

By a process of linking or "chaining" the rules together, a reasoning strategy can be formed. When the rules are linked together by the ES in such a way as to arrive at an action by following a series of conditions and information given as input, it is said to be in the forward-chaining format. For the example shown below (Crowe and Mutch, 1990), the input information on the presence and concentration of Aldicarb in the groundwater is used to arrive at the conclusion:

IF Aldicarb is detected in groundwater
THEN test its concentration
IF concentration is >9 microgram μg/l
THEN groundwater is hazardous
IF the groundwater is hazardous
THEN take remedial action

Rules can also be linked together so as to reach the initial conditions, given the hypothesis. The example given below shows whether use of Aldicarb nearby will contaminate a certain well, by confirming a set of criteria that lead to contamination (Crowe and Mutch, 1990):

IF the well is shallow AND
IF the soil is permeable AND
IF the groundwater flow is toward the well AND
IF the groundwater temperature is about 10°C and pH is 5 to 6
THEN the well probably will be contaminated with Aldicarb

The structure of the ES EXPRES is shown in Figure 3 (Crowe and Mutch, 1990). EXPRES consists of three components:

- Databases
- Inference engine
- User interface

7.5.1. DATABASES

EXPRES has three databases: a knowledge base, a facts base, and an explanations base. These databases have the advantage that they can be updated with additional information as required. The knowledge base contains production rules and, since EXPRES makes use of existing simulation models, the knowledge base also contains numerical models. The knowledge base therefore provides the link between the user and the encoded knowledge, and also undertakes the simulation process through the model. The production rules check for plausible values and relationships among chosen parameters, construct the input data set for the simulation model, and interpret the results of a simulation.

The explanations database consists of a set of encoded explanations provided by an expert that can aid the user in his choice of parameters required as input for a simulation. The explanations database includes definitions and explanations, along with tutorial information about the input parameters. Examples of similar situations and the recommended values that can be input are also included.

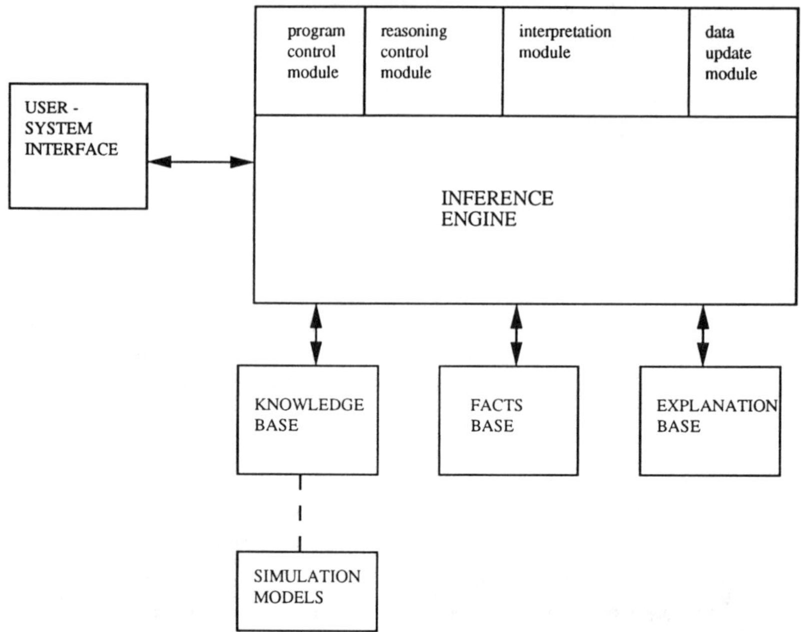

FIGURE 3. Architecture of the EXPRES expert system. (From Crowe, A. S. and Mutch, J. P., *Water Pollut. Res. J. Can.*, 25(3), XX, 1990. With permission.)

The facts database consists of detailed information that describes the features of typical agricultural zones (climatic, hydrogeological, etc.) and knowledge concerning the chemical characteristics of the pesticides used. This knowledge is useful in the event that information about a new pesticide is required by the model. The user can then approximately provide the required data by looking at a family of similar pesticides in the facts database.

7.5.2. INFERENCE ENGINE

The inference engine contains programs that control the ES. Operations such as linking the pesticide transport model, output of the simulation results, etc. are performed. The reasoning strategy and the evaluation of the results provided by the simulation model are also carried out. An interpretation module program within the inference engine translates the user responses to the queries of EXPRES into the required inputs to the simulation model. The entered values are also checked for consistency, and the results of the simulation converted into an easily interpretable form. Through the data update module programs of the inference engine, the data contained in the knowledge base, facts base, or explanations base can be modified or updated upon the receipt of new information.

7.5.3. THE USER INTERFACE

The user interface guides the user through the process of providing input data to the pesticide transport model within EXPRES. Output is conveyed to the user in a standard form of tables and numbers. Graphics are included to help visualize trends and relationships among variables.

7.6. AN EXPERT SYSTEM FOR RESOLUTION OF INDOOR AIR QUALITY (Shoom and Bowen, 1991)

Buildings in recent years depend increasingly on mechanical systems for the circulation and ventilation of fresh air. Energy-saving measures such as sealed windows, greater insulation, and low fresh air openings dictate that natural air exchange with the outdoors should be reduced. The combination of an indoor air pollution source (e.g., materials made with volatile chemicals) and insufficient ventilation to remove the pollutant creates the problem of poor indoor air quality, which is being increasingly encountered. The poor quality of air in a building directly affects the occupants. Even with low levels of air pollution, alertness and the ability to concentrate are impaired. The general comfort and productivity of the occupants are affected, and very poor indoor air quality of a building may even have an impact on the health of the occupants. Since poor air quality can result from widely differing factors such as mechanical system weakness, chemical offgassing, biological contamination, fuel combustion, etc., resolving this problem is quite complex. One needs to identify the shortcomings of ventilation systems and the different pollution sources within the building, and provide a solution. This requires knowledge from varied disciplines, such as chemistry, biology, mechanical engineering, etc. An ES called Indoor Air Quality testing and evaluation Expert System (IAQES) is discussed here. This system was designed to diagnose sources of air pollution and provide possible solutions to remedy the commonly encountered air pollution complaints that occur in office buildings.

A set of testing and measuring instruments, to detect a pollutant source, together with a procedural manual, forms a "test kit" designed for use by personnel evaluating the indoor air quality of a building (Shoom and Bowen, 1991). The manual contains a set of questions, in the form of a "checklist", that an investigator would seek to satisfy through analysis and testing. It also contains interpretations for the answers, and instructions for use of the data acquisition and measuring instruments of the test kit. This manual serves as a basis for the knowledge expertise and rule format of IAQES. The responses given by the user to the questions from the building checklist are interpreted by the system in order to identify the sources and locations of possible pollution problems. This first step forms the data acquisition stage. IAQES then recommends that test procedures be carried out, and makes suggestions about the test instruments to be used, when and where to take the sample, and instrument usage directions. It also provides an interpretation of the test results. This phase is termed the hypothesis formation stage. The most likely solutions are then presented to the user for his consideration, in the final stage.

7.6.1. INFERENCE ENGINE AND KNOWLEDGE REPRESENTATION

The production rules in IAQES follow the backward-chaining format, i.e., the inference engine tries to prove the conclusion by proving its antecedents. Production rules in this ES contain both domain knowledge and flow of control information. The rule structure is in the form of a directed acyclic network, with the rules forming the nodes. The links are formed by the condition-action relationships. Rules closer to the root of the network, or lower-level rules, make up the control rules. They implement the procedural tasks that make up the main menu and menu functions. Rules at higher levels embody the expertise. The rules take the general form

IF <antecedent 1> AND <antecedent 2> AND . . . <antecedent N> THEN <conclusion>

Through this representation of knowledge and rules, the inference engine interprets the user's checklist answers and forms hypotheses about the nature of the problem.

The Preliminary Assessment Building Checklist (PABC) is one of several logical blocks or modules of this ES. This module is usually the first one referred to by the inference logic during an investigation. It asks for answers from the building checklist questions, and thereby determines possible air quality problems, their sources and locations, and where the user should test.

The following types of indoor air quality problems can be detected through IAQES:

- Carbon monoxide and other combustion byproducts
- Biological contamination
- Volatile organic compounds
- Formaldehyde
- Airborne particulates
- HVAC and mechanical system problems
- Temperature and humidity
- Radon

7.6.2. FINDING POSSIBLE SOURCES AND SUBSOURCES

Each of the above air quality problems can stem from several possible sources, or specific causes, of pollution. In the case of a building with an indoor loading dock, for example, the loading dock is a possible source of carbon monoxide pollution. IAQES views many sources of pollution along with their "subsources". Subsources are factors that provide evidence for a possible source. In the case of the building with an indoor loading dock, the fact that the doors of the dock close after a truck enters the dock strengthens evidence that the dock may be a carbon monoxide source. Thus, if none of the possible subsources of a given source are present, the probability of its being the cause of pollution is low. This probability is highest when all subsources are present. The subsources are determined directly from answers to questions posed to the user by the building checklist. A source is determined either directly from the answers to single questions or from a group of subsources. Sources are viewed as weighted or unweighted. A weighted source is one that has a value associated with it, indicating the likelihood of a problem originating from the source. A weighted source is determined from the presence of its subsources. The weight, or probability of occurrence of a source is determined from the ratio:

$$\frac{\text{Proportion of source's subsources}}{\text{Total number of subsources present}}$$

Unweighted sources are those that usually lack subsources, and depend on the answer to a single question. They are assigned weights of 1 if present, and 0 if absent.

Possible problems of indoor air quality are inferred from the presence of sources. When IAQES finds a possible source, it assigns a value of 1 to a flag, in a list (called the possible problem list) that contains all the indoor air quality problems for which evidence has been found. For example, consider a building whose mechanical ventilation system has wet and dirty air intake dust filters. Microbes could flourish in, and airborne dirt particles be released from such filters. Hence, the system would assign 1 to each of two flags, "biological score" and "particular score", and add both biological contamination and presence of airborne particulates to the possible problem list for the user to consider.

Possible problem locations are determined by the system, based on user responses, nature of the possible sources, and rules of thumb acquired from the expert.

After testing for pollutants, the user could enter the readings and use the ES to provide an interpretation, and then a summary describing the measure and implications of the results.

The solutions offered by IAQES are both general and specific in nature. For example, in the case of biological contamination, the ES would recommend that the infected places be thoroughly disinfected and the wet and dirty air filters replaced by new ones.

The various functional modules of IAQES are shown in Figure 4 (Shoom and Bowen, 1991). The logical blocks of Figure 4 implement the various

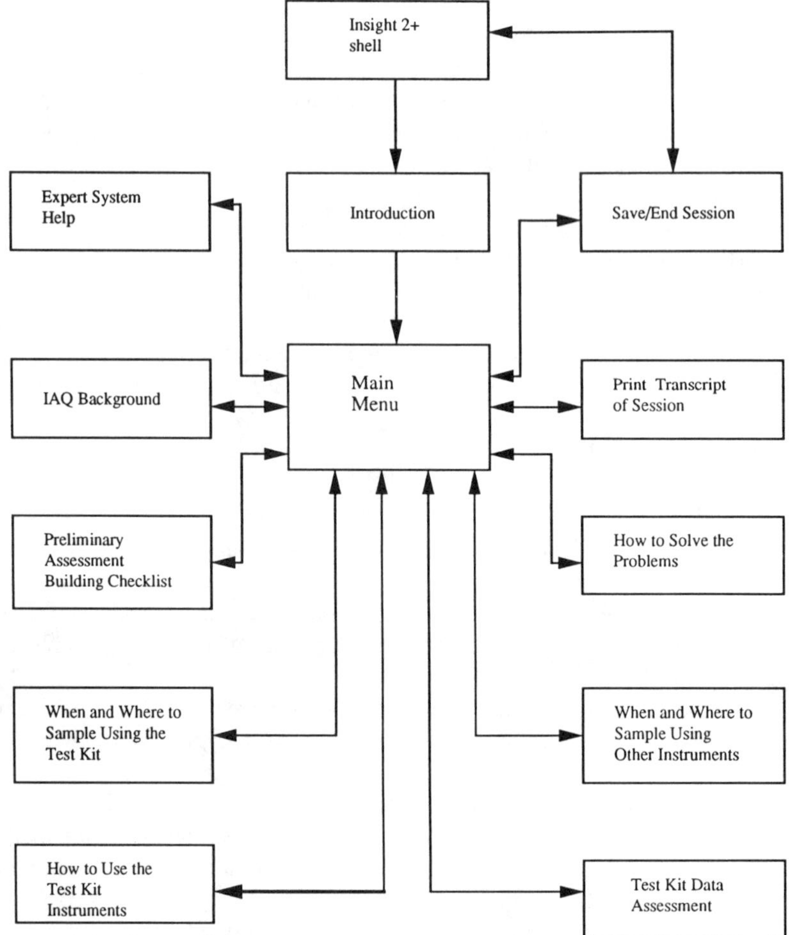

FIGURE 4. IAQES logical structure. (From Shoom, G. B. and Bowen, D. G., *Proc. 5th Annu. AI Systems in Govt. Conf.*, IEEE Service Center, Piscataway, NJ, 1991. With permission.)

functions of data gathering, hypothesis formation, test recommendation, and provision of solutions. Each block, or module, performs one main task, and some diagnostic functions map onto several blocks.

Introduction—This module displays an introductory text for a user beginning a new session, or allows the user to continue an existing investigation, by restoring information from an earlier session.

Main Menu—This module allows the user to invoke the various functions of the ES.

System Help—On-line help can be requested at any point during the session through this module.

When and Where to Use Sample Kit—This functional block provides knowledge about (1) what instrument(s) are to be used, (2) when to sample, (3) where to control the sample, and (4) possible sources found by the PABC and their locations.

How to Use the Instruments—This module provides usage instructions.

Data Assessment—The test results for a list of indoor air quality problems are read in and interpreted by this module. For example, indoor levels of carbon monoxide between 4 and 9 ppm may cause complaints about air quality, and levels above 9 ppm are considered dangerous and reported as such by the system.

Where and When to Sample Using Other Instruments—This module proposes tests that use instruments not available in the test kit.

Print Session Transcript—The transcript of the session is automatically compiled by the system as the user works, and sent to a printer.

7.6.3. SYSTEM DEVELOPMENT

IAQES was developed using the ES development shell INSIGHT2+ for use on a personal computer under MS-DOS. INSIGHT2+ production-rule based, with the rules following a backward-chaining format. Further modifications to the system can be carried out and more indoor air quality problems included as part of the system upgrading process.

7.7. A DECISION SUPPORT SYSTEM FOR RESERVOIR ANALYSIS (Slobodan et al., 1989)

Reservoirs constitute one of the most important water resource systems. They are used for flood control, water supply, low flow augmentation, hydroelectric power applications, etc. The adoption of systems analysis techniques for the planning, design, and management of complex water resource systems constitutes an important advance in reservoir studies. However, analytical models possess shortcomings, in that water resource experts feel that their experience and understanding of the system are not fully incorporated into the models. The models are not tested on a variety of real-life situations, and the information exchange between the modeler and the water resource engineer could be unclear. Expert systems (ES) have proved highly applicable to the field of water resource engineering. Their applications include selection of reservoir sites, determination of reservoir size, evaluation of reservoir design characteristics, monitoring of reservoir operation, etc. The application of ES technology to the problem of reservoir management provides several advantages: a saving of the expert's time, increased understanding of a particular reservoir problem, and useful training capability for the engineer. An ES designed for use as an advisory tool for reservoir analysis (called REZES) is discussed here.

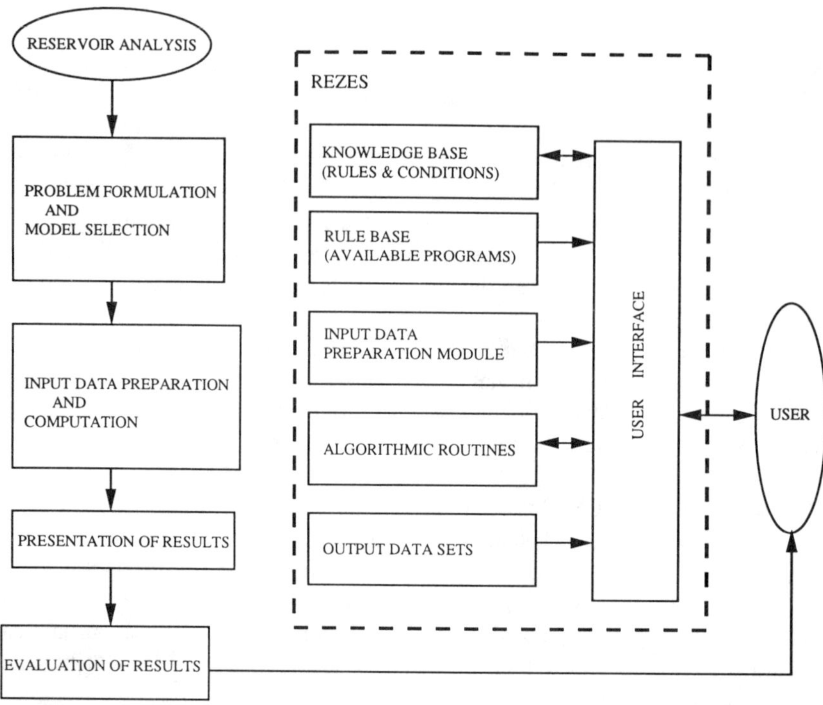

FIGURE 5. REZES architecture. (Simonovic, S. P. and Savic, D. A., *J. Comp. Civil Eng.*, 3(4), 367, 1989. With permission.)

The architecture of the REZES is shown in Figure 5 (Slobodan et al., 1989). The major components of this ES are:

- Knowledge base
- Program library
- Input data preparation module
- Algorithmic routines
- User interface

7.7.1. KNOWLEDGE BASE

Knowledge representation in REZES is rule based. The knowledge base contains the production rules and conditions of the ES. Rules are of the IF <condition> THEN <action> format. An example of a rule is given below (Slobodan et al., 1989):

IF reservoir exists
AND detailed analysis is required
AND length of inflow record is sufficient

OR inflow probability distribution functions are available
THEN stochastic model should be used

As seen, the rules follow a backward-chaining format, with the conclusion preceding the conditions.

7.7.2. PROGRAM LIBRARY

REZES contains five reservoir models that form the library for the system, as given below (Slobodan et al., 1989):

1. RESER: a deterministic simulation-optimization model for sizing a multipurpose reservoir storage. This model is used by arithmetic routines to determine the size of conservation storage that satisfies reliability and vulnerability criteria.
2. ILP: an iterative linear programming model for planning the monthly reservoir hydropower production over the year. This model is used by the arithmetic routines; it maximizes the value of the hydroenergy generated and the expected future returns represented by water left in the reservoir.
3. CCCP: a multipurpose reservoir chance-constrained model for planning the monthly reservoir operation over the year. This model is used to find the optimal release policy.
4. RPORC: a stochastic reliability programming model for planning the monthly operation of a multipurpose reservoir. The arithmetic routines use this algorithm for optimization of the reservoir releases.
5. PROFEXY: a real-time multipurpose reservoir operation model with a daily time step.

7.7.3. INPUT DATA PREPARATION MODULE

After selection of a mathematical model from the program library, the ES uses the input data preparation module to provide explanations, examples, and the correct format for the input variable.

7.7.4. ARITHMETIC ROUTINES

These routines use the models in the program library to determine various parameters related to the reservoir design problem.

7.7.5. USER INTERFACE

This module is the interface between the user and the ES. Through the user interface, the user provides the system with the necessary information about a problem under analysis. The system uses these inputs to select an appropriate model for solving the problem, and displays the results to the user through the output data representation module. The output data representation module presents the results of the session for evaluation by the user.

Use of REZES starts with the menu screen, which offers a menu from which to select various groups of system actions.

REZES was programmed, using the language PROLOG, as a rule-based structure. REZES was created by experts in the field of reservoir management. The system was designed to help users in problem formulation and model selection, input data preparation and computation, and presentation and evaluation of results.

REFERENCES

AASHTO, *Standard Specifications for Highway Bridges,* 13th ed., American Association of State Highway and Transportation Officials, Washington, D.C., 1983.

Adeli, H., *Expert Systems in Construction and Structural Engineering,* Chapman and Hall, New York, 1988.

Adeli, H. and Balasubramanyam, K. V., A knowledge-based system for design of bridge trusses, *J. Comput. Civ. Eng.,* 2(1), 1, 1988.

AISC, *Manual of Steel Construction,* 8th ed., American Institute of Steel Construction, Chicago, 1978.

Allwood, R. J., Stewart, D. J., and Trimble, E. G., Some experiences from evaluating expert system shell programs and some potential applications, in *Proc. 2nd Int. Conf. Civil and Structural Engineering Computing,* Vol. 2, Civil-Comp. Press, London, 1985, 415.

Arockiasamy, M., Sawka, M. J., Sinha, V., and Shahawy, M., Knowledge based expert system for rating and assessment of Florida bridges, Second Int. Conf. on Bridge Management, University of Surrey, Guilford, Surrey, U.K., April, 1993.

Arockiasamy, M. and Sunghoon Lee, State-of-the-art on expert systems applications in design construction and maintenance of structures, Contract Report ITL-89-1, Computer-Aided Structural Engineering (CASE) Project, U.S. Army Engineer Waterways Experiment Station, Vicksburg, MS.

Ashley, D. B. and Wharry, M. B., Prototype expert system for subsurface risk, National Science Foundation Grant CEE-8352354, 1985.

Bennett, J. S. and Englemore, R. S., SACON: a knowledge based consultant for structural analysis, in Proc. 6th Int. Joint Conf. Artificial Intelligence, Tokyo, 1979, 47.

Biegler, L. T. and Cuthrell, J. E., Improved infeasible path optimization for sequential modular simulators: the optimization algorithm, *Comput. Chem. Eng.,* 9(1), 257, 1985.

BOCA, *The BOCA Basic Building Code,* 9th ed., Building Officials and Code Administrators International, Country Club Hills, IL, 1984.

Bowen, J., Cornick, T. C., and Bull, S. P., BERT—an expert system for brickwork design, working paper, University of Reading, England, Departments of Computer Science and Construction Management, 1986.

Burnham, G., Gaskell, A., Hutchinson, P., and White, P., Knowledge base for an expert system—the unified soil classification system, Computer Applications Research Unit, Department of Architectural Science, University of Sidney, unpublished.

Camacho, G., LOW-RISE: An expert system for structural planning and design of industrial buildings, M.S. thesis, Carnegie Mellon University, Department of Civil Engineering, Pittsburgh, PA, 1985.

Carter, M., *Geotechnical Handbook,* Pentech Press, Plymouth, 1983.

Crowe, A. S. and Mutch, J. P., Assessing the migration and transformation of pesticides in the subsurface, *Water Pollut. Res. J. Can.,* 25(3), 000, 1990.

De La Garza, J. M. and Ibbs, C. W., Issues in construction scheduling knowledge presentation, in *Managing Construction Worldwise,* Vol. 1, *Systems for Managing Construction,* Lansley, P. R. and Harlow, P. R., Eds., E. & F. N. Spon, New York, 1987, 543.

Diaz, M. A., Carmichael, R. F., III, Hudson, S. W., Moser, L. O., and Grant, M., A user-friendly data base module for bridge management systems, in *Developments in Short and Medium Span Bridge Engineering '90,* Toronto, August 1990, 63.

Faghri, A. and Demetsky, M. J., Expert system for traffic control in work zones, *J. Transp. Eng.,* 116(6), 000, 1990.

Fenske, T. E. and Fenske, S. M., Expert systems for highway bridge analysis and design, in *Developments in Short and Medium Span Bridge Engineering '90,* Toronto, August 1990, 23.

Fenves, S. J. and Norabhoompipat, T., Potential for artificial intelligence applications in structural design and detailing, in *Artificial Intelligence and Pattern Recognition in Computer-Aided Design,* Latombe, J. D., Ed., Elsevier/North Holland, Amsterdam, 1978.

Fenves, S. J., Software for analysis of standards, in Proc. 2nd ASCE Conf. Computing in Civil Engineering, Baltimore, June 1980, 82.

Fenves, S. J., Maher, M. L., and Sriram, D., Expert systems: C.E. potential, in Civil Engineering/ASCE, October 1984, 45.

Fenves, S. J., What is an expert system?, in Proc. ASCE Symp. Expert Systems in Civil Engineering, Kostem, C. N. and Maher, M. L., Eds., Seattle, 1986, 1.

Fenves, S. J., Flemming, U., Hendrickson, C., Maher, M. L., and Schmiddt, G., Integrated Software Environment for Building Design and Construction, *Computer Aided Design,* Vol. 22, No. 1, Butterworth, Guildford, England, 1990, 27.

Finn, G. A. and Reinschmidt, K. F., Expert system in an engineering-construction firm, in Proc. ASCE Symp. Expert Systems in Civil Engineering, Kostem, C. N. and Maher, M. L., Eds., Seattle, 1986, 40.

Firlej, M., Training and education for knowledge engineers, in Proc. 1st Int. Expert Systems Conf., London, October 1 to 3, 1985, 209.

Forgy, C. L., *OPS5 User's Manual,* Tech. Rep. CMU-CS-81–135, Carnegie-Mellon University, Pittsburgh, PA, 1981.

Garrett, J. H., Jr., SPEX—A Knowledge-Based Standard Processor for Structural Component Design, Ph.D. thesis, Carnegie-Mellon University, Pittsburgh, PA, 1986.

Gaschnig, J., Reboh, R., and Reiter, J., *Development of Knowledge-Based Systems for Water Resources Problems,* SRI Proj. 1619, Stanford Research Institute, Menlo Park, CA, 1981.

Gero, J. S. and Coyne, R. D., Developments in expert systems for design synthesis, in *Expert Systems in Civil Engineering,* American Society of Civil Engineers, New York, 1986.

Hamiani, A. and Popesen, C., CONSITE: a knowledge based expert system for site layout, in Proc. 5th ASCE Conf. Computing in Civil Engineering: Microcomputers, Will, K. M., Ed., 1988, 248.

Harris, J. R. and Wright, R. N., Computer aids for the organization of standards, in Proc. 2nd ASCE Conf. Computing in Civil Engineering, Baltimore, June 1980, 92.

Hays, C. O., Jr., Hoit, M. I., Selvappalam, M., and Vinayagamonthy, M., Development of automated bridge rating program using finite element technology, in *Developments in Short and Medium Span Bridge Engineering '90,* Toronto, August 1990, 11.

Hendrickson, C. T., Zozaya-Gorostiza, C. A., Rehak, D., Baracco-Miller, E. G., and Lim, P. S., An expert systems for construction planning, *ASCE J. Comput.,* October 1987.

Howard, H. C., Interfacing Databases and Knowledge Based Systems for Structural Engineering Applications, Ph.D. thesis, Carnegie-Mellon University, Pittsburgh, PA, 1986.

Howell, T. F., Automation for transportation—more than data processing, *J. Transp. Eng.,* 116(6), 000, 1990.

Hutchinson, P., An Expert System for the Selection of Earth Retaining Structures, M.S. thesis, University of Sydney, Department of Architectural Science, Australia, 1985.

IBM, *Graphical Data Display Manager: Application Programming Guide,* Program 5748-XXH, Release 4, 3rd ed., IBM Corporation, Cary, NC, 1984.

Knowles, L., Henry, J. P., and Shafer, M., Expert system for evaluating and notifying hazardous generators, *J. Comput. Civ. Eng.,* 3(2), 000, 1989.

Kostem, C. N., Design of an expert system for the rating of highway bridges, in *Expert Systems in Civil Engineering,* American Society of Civil Engineers, New York, 1986.

Kunz, J. C., Bonura, T., and Stelzner, M. J., Applications of artificial intelligence in engineering problems, in *Proc. First International Conference, Southampton University, U.K.,* Vol. 2, Sriram, D. and Adey, R. A., Eds., Springer-Verlag, New York, 1986.

Levitt, R. E., HOWSAFE: a microcomputer-based expert system to evaluate the safety of a construction firm, in Proc. ASCE Symp. Expert Systems in Civil Engineering, Kostem, C. N. and Maher, M. L., Eds., Seattle, 1986, 55.

Lin, T. Y. and Stotesbury, S. D., *Structural Concepts and Systems For Architects and Engineers,* John Wiley & Sons, New York, 1981.

Maher, M. L., HI-RISE: A Knowledge-Based Expert System for the Preliminary Design of High Rise Buildings, Ph.D. thesis, Carnegie-Mellon University, Pittsburgh, PA, 1984.

Maher, M. L., Problem solving using expert system techniques, in *Expert Systems in Civil Engineering,* American Society of Civil Engineers, New York, 1986, 7.

Maher, M. L., *Expert System Components,* ASCE Expert Systems Comm. Rep. Expert Systems for Civil Engineers: Technology and Application, American Society of Civil Engineers, New York, 1987.

Needham, T. and Andersen, N. H., Experience in the development of a bridge classification system and implementation of an integrated design and analysis system, in *Developments in Short and Medium Span Bridge Engineering '90,* Toronto, August 1990, 351.

Nguyan, R. P., Development of prototype expert systems for civil engineering design, in Proc. ASEE Annu. Conf., 1990, 1856.

Niwa, K. and Okuma, M., Know-how transfer method and its application to risk management for large construction projects, *IEEE Trans. Eng. Manage.,* 29(4), 146, 1982.

O'Connor, M. J., De La Garza, J. M., and Ibbs, C. W., Jr., An expert system for construction schedule analysis, in Proc. ASCE Symp. Expert Systems in Civil Engineering, Kostem, C. N. and Maher, M. L., Eds., Seattle, 1986, 67.

Orhun, E. and Demirars, O., A PC+ based oil pollution combating system, in *14th Annual Energy-Sources Technology Conference and Exhibition,* ASME, 1991.

Ovunc, B. A., Knowledge-based expert system for tapered frames, in *Computing in Civil Engineering,* ASCE, New York, 1989, 28.

Paulson, B. C., Jr. and Sotoodeh-Khoo, H., Expert systems in real-time construction operations, in *Managing Construction Worldwide,* Vol. 1, *Systems for Managing Construction,* Lansley, P. R. and Harlow, P. A., Eds., E. & F. N. Spon, New York, 1987.

PENNDOT, *Bridge Analysis and Rating BARG,* User's Manual for Computer Program P4353110, PENNDOT, March 1989.

PCI, *Precast Prestressed Concrete Short Span Bridges Spans to 100 ft,* Prestressed Concrete Institute, Chicago, 1975.

Rasdorf, W. J. and Fenves, S. J., Design specifications representation and analysis, in Proc. 2nd ASCE Conf. Computing in Civil Engineering, Baltimore, June 1980, 102.

Rasdorf, W. J. and Wang, T., Expert system integrity maintenance for the retrieval of data from engineering databases, in Proc. 4th ASCE Conf. Computing in Civil Engineering, Boston, October 1986, 654.

Rasdorf, W. J. and Wang, T. E., Generic design standards processing in an expert system environment, *J. Comput. Civ. Eng.,* 2(1), 68, 1988.

Rathi, A. K., Santiago, A. J., Valentine, D. E., and Chin, S. M., ITDS: past, present, and future, *J. Transp. Eng.,* 116(6), 000, 1990.

Reboh, R., *Knowledge Engineering Techniques and Tools in the PROSPECTOR Environment,* SRI Tech. Note 243, Stanford Research Institute, Menlo Park, CA, 1981.

Reddy, D. R., Erman, L. D., and Neely, R. B., The HEARSAY speech understanding system: an example of the recognition process, in Proc. Int. Joint Conf. AI, Tokyo, 1973, 185.

Rehak, D. R. and Fenves, S. J., Expert systems in construction, in Proc. 1984 ASME Int. Computers in Engineering Conference and Exhibit, Vol. 1, Las Vegas, Gruver, W. A., Ed., 1984, 228.

Rooney, M. and Smith, S. E., Artificial intelligence in engineering design, *Comput. Struct.,* 16, 279, 1983.

Roschke, P. N., Advisory system for design of highway safety structures, *J. Transp. Eng.,* 117(4), 000, 1991.

Rosenmann, M. A., Gero, J. S., and Oxman, R., An expert system for design codes and design rules, Applications of AI to Engineering Problems, 1986.

Russell, J. S. and Skibniewski, M. J., An expert system for contractor prequalification, in Proc. 5th ASCE Conf. Computing in Civil Engineering: Microcomputers to Supercomputers, Will, K. M., Ed., 1988, 239.

Saouma, V. E., Jones, M. S., and Doshi, S. M., A PC based expert system for automated reinforced concrete design checking, Rep. Contract DACA39–86–K–0011, U.S. Army Engineer Waterways Experiment Center, Information Technology Laboratory, Vicksburg, MS, 1987.

Senerviratne, P. N., Knowledge-based system for managing intersection safety, *J. Transp. Eng.*, 116(6), 000, 1990.

Shoom, G. B. and Bowen, D. G., IAQES: an expert system for indoor air quality problem resolution, in Proc. 5th Annu. AI Systems in Government Conf., IEEE Service Center, Piscataway, NJ, 1991.

Slobodan, P., Simonovic, and Savic, D. A., Intelligent decision support and reservoir management and operations, *J. Comput. Civ. Eng.*, 3(4), 1989.

Stefik, M., Planning with constraints, Tech. Rep. STAN-CS-80–784, Stanford University, Computer Science Department, 1980.

Teknowledge, *M.1 Training Materials, M.1 Reference Manual,* Version 1.3, Teknowledge, Inc., Palo Alto, CA, 1985.

Terzahgi, K. and Peck, R. B., *Soil Mechanics in Engineering Practice,* John Wiley & Sons, New York, 1967.

Tommelein, I. D., Levitt, R. E., and Hayes-Roth, B., Using expert systems for the layout of temporary facilities on construction sites, in *Managing Construction Worldwide,* Vol. 1, *Systems for Managing Construction,* Lansley, P. R. and Harlow, P. A., Eds., E. & F. N. Spon, New York, 1987, 566.

VanMelle, W., A domain independent production-rule system for consultation programs, in Proc. 6th Int. Joint Conf. Artificial Intelligence, Tokyo, August 1979, 923.

VanMelle, W., A domain-dependent production rule system for consultation programs, in Proc. 6th Int. Joint Conf. Artificial Intelligence, Tokyo, August 1979, 923.

Wang, T. and Rasdorf, W. J., *Query Watcher User's Guide,* Version 1.0, Computer Studies Program, North Carolina State University, Raleigh, 1985.

Wang, T., *Generic Design Standards Processing in a Knowledge-Based Expert System Environment,* M.S. thesis, North Carolina State University, Raleigh, 1986.

Wang, T. and Rasdorf, W. J., *SPIKE User's Guide, Version 1.0,* Computer Studies Program, North Carolina State University, Raleigh, 1986.

Welch, J. and Biswas, M., Application of expert systems in the design of bridges, Tech. Rep., 1986.

Wilson, F. R., Cantin, L., and Bisson, B. G., Emergency response system for dangerous goods movement by highways, *J. Transp. Eng.*, 116(6), 000, 1990.

Wilson, J. L., Mikroudis, G. K., and Fang, H. Y., GEOTOX: a knowledge-based system for hazardous site evaluation, *Artif. Intell.*, 2(1), 000, 1987.

Zhou, H. and Layton, R. D., Development of prototype expert system for roadside safety, *J. Transp. Eng.*, 117(4), 000, 1991.

Zozaya-Gorostiza, C., Hendrickson, C., Baracco, E., and Lim, P., *Decisions Tables for Knowledge Representation and Acquisition,* Tech. Rep., Carnegie Mellon University, Department of Civil Engineering, Pittsburgh, PA, 1987.

INDEX